The Domestic Politics of Global Climate Change

NEW HORIZONS IN ENVIRONMENTAL POLITICS

Series Editor: Arthur Mol, *Chair and Professor in Environmental Policy, Director, Wageningen School of Social Sciences, Wageningen University, the Netherlands, Professor in Environmental Policy, Renmin University, Beijing*

The New Horizons in Environmental Politics series provides a platform for in-depth critical assessments of how we understand the many changes in the politics of nature, the environment and natural resources that have occurred over the last 50 years. Books in the series question how the environment is (re)defined, debated and protected; explore differences between countries and regions in environmental politics; analyse how actors do and do not collaborate around environment and natural resource conflicts; describe who wins and who loses and in what ways; and detail how to better study, analyse and theorize such developments and outcomes.

The series is designed to promote innovative cross-disciplinary analysis of the contemporary issues and debates influencing the various dimensions of environmental politics. Covering a diverse range of topics, the series will examine the political, economic and ethical aspects of environmental policy, governance and regulation. It brings together cutting edge research on environmental politics worldwide in order to shed light on and explain current trends and developments.

With oversight from the Series Editor, Professor Arthur Mol – a noted specialist in the field of environmental politics at Wageningen University, the Netherlands – the New Horizons in Environmental Politics series comprises carefully commissioned projects from experts in the field including both academics and professionals. The audience for the series is global, and books in the series are essential reading for students, academics and professionals – in short, anyone with an interest in understanding the vital issues affecting environmental politics in the twenty-first century.

Titles in the series include:

Russia and the Politics of International Environmental Regimes
Environmental Encounters or Foreign Policy?
Anna Korppoo, Nina Tynkkynen and Geir Hønneland

American Environmental Policy
The Failures of Compliance, Abatement and Mitigation
Daniel Press

Environmental Governance through Partnerships
A Discourse Theoretical Study
Ayşem Mert

The Domestic Politics of Global Climate Change
Key Actors in International Climate Cooperation
Edited by Guri Bang, Arild Underdal and Steinar Andresen

The Domestic Politics of Global Climate Change

Key Actors in International Climate Cooperation

Edited by

Guri Bang

Center for International Climate and Environmental Research – Oslo (CICERO), Norway

Arild Underdal

Department of Political Science, University of Oslo and Center for International Climate and Environmental Research – Oslo (CICERO), Norway

Steinar Andresen

Fridtjof Nansen Institute, Norway

NEW HORIZONS IN ENVIRONMENTAL POLITICS

Edward Elgar
PUBLISHING

Cheltenham, UK • Northampton, MA, USA

Published by
Edward Elgar Publishing Limited
The Lypiatts
15 Lansdown Road
Cheltenham
Glos GL50 2JA
UK

Edward Elgar Publishing, Inc.
William Pratt House
9 Dewey Court
Northampton
Massachusetts 01060
USA

A catalogue record for this book
is available from the British Library

Library of Congress Control Number: 2015938868

This book is available electronically in the **Elgar**online
Social and Political Science subject collection
DOI 10.4337/9781784714932

MIX
Paper from responsible sources
FSC® C013056
FSC www.fsc.org

ISBN 978 1 78471 492 5 (cased)
ISBN 978 1 78471 493 2 (eBook)

Typeset by Columns Design XML Ltd, Reading
Printed and bound in Great Britain by TJ International Ltd, Padstow

Contents

Contributors

Solveig Aamodt, Center for International Climate and Environmental Research – Oslo (CICERO), Norway

Steinar Andresen, Fridtjof Nansen Institute, Norway

Guri Bang, Center for International Climate and Environmental Research – Oslo (CICERO), Norway

Masahiko Iguchi, Department of International Relations, Kyoto Sangyo University, Japan

Anna Korppoo, Fridtjof Nansen Institute, Norway and Academy of Finland, Postdoctoral Fellow, Turku University, Finland

Alexandru Luta, Sandbag Climate Campaign, United Kingdom

Trude Rauken, Center for International Climate and Environmental Research – Oslo (CICERO), Norway

Jon Birger Skjærseth, Fridtjof Nansen Institute, Norway

Iselin Stensdal, Fridtjof Nansen Institute, Norway

Sunil Tankha, International Institute of Social Studies, Erasmus University Rotterdam, the Netherlands

Arild Underdal, Department of Political Science, University of Oslo and Center for International Climate and Environmental Research – Oslo (CICERO), Norway

Preface

Why are some countries more willing and able than others to engage in climate change mitigation? This book compiles the insights of experts in comparative politics and international relations to analyse the policy trajectories of seven key actors in the international climate negotiations: Brazil, China, the European Union, India, Japan, Russia and the United States. All these actors are among the world's largest greenhouse gas emitters, together accounting for nearly 70 per cent of total world emissions. Moreover, to varying degrees they all speak on behalf of larger groups in the global climate change negotiations. Combining these two observations, we may consider each of these actors, with Japan and Russia as likely exceptions, capable of blocking or at least delaying a truly effective global mitigation agreement – some by their own weight, others with the support of 'their' group. The good news is that any agreement signed and ratified by all seven will almost certainly be accepted by an overwhelming majority of other states. Overall, the outcome of the United Nations climate change negotiations critically depends on the willingness and ability of our sample of actors to contribute to mitigating greenhouse gas emissions.

We proceed as follows. First, we develop an analytical framework designed to map and explain similarities and differences in domestic policies as well as in the roles these actors play in the global climate change negotiations. Second, the authors perform in-depth case studies analysing the climate policy trajectories of the seven key actors. Third, a comparative analysis of the main drivers and barriers that shape climate and energy policy is undertaken and discussed in the context of mega-trends like population growth, affluence and technology development. The contributing authors are participants in a common research project – Strategic Challenges in International Climate and Energy Policy (CICEP) – that joins the forces of CICERO (the Center for Climate and Environmental Research – Oslo), the Fridtjof Nansen Institute and the University of Oslo as a Centre for Environment-Friendly Energy Research (FME) to study crucial questions related to international climate collaboration. The authors are leading experts on climate policy development in the seven key actors and on international climate agreements, and the book provides updated knowledge as well as theoretical perspectives that should be of interest to the international research community studying global environmental politics and international climate negotiations, to students of environmental politics, to practitioners in the energy sector, and for policy-makers involved in climate and energy policy decision processes.

We would like to thank the Research Council of Norway and our user partners for generous funding of the CICEP centre (project number 209701), which made this book possible. We also thank several colleagues for their time and effort in commenting on individual chapters, in particular Jon Hovi, Eduardo Viola, Jost Wübbeke, Olav Schram Stokke, Joyeeta Gupta, Norichika Kanie, Arild Moe and Tora Skodvin. Lan Marie Nguyen Berg provided useful assistance with database development. Last but not least, we thank Frank Azevedo for excellent editorial assistance.

Guri Bang, Arild Underdal and Steinar Andresen

1. Introduction

Guri Bang, Arild Underdal and Steinar Andresen*

PURPOSE AND SCOPE

Why are some countries more willing and able than others to engage in climate change mitigation? This book's authors address that question by exploring climate and energy policy trajectories of seven key actors that play crucial and different roles in global climate cooperation: Brazil, China, the European Union, India, Japan, Russia and the United States. We map similarities and differences in domestic policies as well as in the roles these actors play in the global climate change negotiations. Moreover, to understand these similarities and differences we undertake a comparative analysis of the main drivers and barriers that shape climate and energy policy trajectories in each case.

We have three reasons for focusing this project on developments at the level of major countries and regions. First, even though international cooperation will be required to effectively mitigate the impact of human activities on the global climate system, domestic policies and measures constitute the basic building blocks of any international climate agreement. Decisions made and strategies adopted at the national, regional and, increasingly, at the subnational level determine the speed and force with which the climate change problem is being addressed. David Victor (2011) has convincingly argued that global conference diplomacy involving more than 190 countries bargaining about distributing obligations to contribute to a global mitigation programme is not likely to arrive at solutions that 'prevent dangerous anthropogenic interference with the climate system' (UNFCCC, Art. 2). His advice is to adopt an alternative approach that 'starts with what nations are willing and able to implement' (Victor, 2011, p. 6) and that 'starts with the interests of the most powerful countries' (ibid., p. 265). Here, we follow that advice.

Interestingly, some of the largest greenhouse gas (GHG) emitters have taken on increasingly direct roles in international climate policy since

2009 (Gupta, 2010; Victor, 2011). Global climate negotiations have shifted significantly; major emitters now prefer to develop the foundations of international climate policy from the bottom up by taking action domestically. This shift from a top-down–oriented regime to a bottom-up approach recognizes the boundaries that countries' domestic politics set for their participation at the international level (Falkner et al., 2010; Andresen, 2015). A telling example is the miniscule role that the Kyoto Protocol plays in its second commitment period, with very few Annex 1 countries with reduction commitments left as members of the agreement. The fundamental shift towards negotiations where countries' special circumstances and individual assessments of politically feasible policy action take centre stage underscores the need for thorough and systematic analysis of key actors in the international climate negotiations. Any new climate agreement must 'ensure that key industrialized and developing nations are involved in differentiated but meaningful ways' (Olmstead and Stavins, 2012, p. 66). In essence, such differentiation requires that policy-makers and negotiators be able to identify and to some extent manipulate the drivers and barriers that determine what each country is willing and able to do.

Second, the combination of increasingly confident Intergovernmental Panel on Climate Change (IPCC) assessments and very slow progress in the UN Framework Convention on Climate Change (UNFCCC) negotiations over the past several years has spurred interest in mitigation measures that can be developed and implemented at lower levels – subnational, national and regional. Although a quintessential global public goods challenge, the human impact on the climate system can be traced back to a very wide range of activities, from fossil fuel burning to land use and land cover change. This multiple source nature of the problem has at least two important implications. The bad news is that, as far as presently known, there can be no quick fix. The good news is that with multiple sources of GHG emissions come multiple opportunities for countries, communities, companies and even individuals to make small but constructive contributions to mitigation. Many of the new initiatives that have been taken over the past five to ten years seem to have emerged through synergistic interplay involving public awareness and concern, technological innovation, growing market demand, and government policies. One case in point is the remarkably rapid growth in solar energy production and consumption. This development has been driven, in large part, by a combination of substantial improvement in solar panel efficiency, much lower equipment prices achieved through large-scale production and growing demand, and government regulations and incentive schemes promoting the transition to low-carbon energy systems. In some instances, China being a prominent example, the rapid increase in renewable energy investments seems to be driven as much (or more) by concern about local pollution and health problems as by concern about damage caused by climate change.

Third, the IPCC's latest assessment report points out that despite some encouraging developments, policies and strategies to reduce GHG emissions are in their early stages in most countries, that too little is known about their effects, and that more studies are needed of how the design of political institutions influences the choice and feasibility of policy options (IPCC, 2014, p. 5). One of the knowledge gaps identified is that 'various jurisdictions produce various policy instruments that are influenced by co-benefits and political economy, and these policy instruments interact in complex manners which makes it difficult to evaluate the economic and environmental effectiveness of individual policy instruments as well as the policy package of a nation' (ibid., p. 10). This volume will help fill the knowledge gap by analysing the drivers for and barriers to climate change mitigation across different institutional settings, political cultures, and power configurations in seven of the most important actors.

Many countries play significant roles in the negotiations, which has as an aim a global treaty accepted by all UN member states. In functional as well as political terms, however, some actors are far more important than others. We see at least three reasons why analyses of key actors in the international climate negotiations should include the seven that we have selected.

First, the selected key actors are among the largest GHG emitters in the world. Together they produce nearly 70 per cent of the world's total GHG emissions; the combined GHG emissions of the United States and China represent the largest share, about 45 per cent (World Resources Institute, 2015). For many years, the United States was the largest emitter but was surpassed by China in 2007. China's emissions have grown exponentially over the past 20 years in step with the country's impressive economic growth. This growth has largely been fuelled by coal-powered electricity; coal is the energy source for nearly 70 per cent of total electricity generation in China (EIA, 2014). In the United States, new technologies for horizontal drilling and hydraulic fracturing have since 2009 allowed massive extraction of natural gas, cutting prices to a level that has led to a switch from coal to natural gas as the most preferred energy source in electricity generation. The reliance on coal for electricity production has dropped from 49 per cent in 2007 to 40 per cent in 2013, resulting in a substantial drop in CO_2 emissions (EIA, 2013). Because of the strong emissions increase in China and the steep reduction in the United States, China now emits almost twice as much CO_2 as the United States.

India and Brazil are also expected to increase their emissions substantially as their economies continue to grow. India increasingly depends on coal for electricity generation, even though it has developed more capacity for wind energy over the past ten years. Rapid population growth and access to electricity for an increasing share of citizens are central to explaining the emissions growth trend in India. In Brazil the discovery of offshore petroleum has paved the way for more use of fossil fuels in its transport sector. Brazil's transport sector has for decades been very clean compared to the transport sectors of other countries because of the large share of biofuels in the mix. Cheap and accessible oil is likely to replace some of the ethanol. Moreover, the success in reducing deforestation has tempered since 2012. For the European Union the trend is different, with emissions levels stabilizing or slightly decreasing because of the combination of lower population growth, economic stagnation and phase-in of more renewable energy. Because of high energy efficiency, Japan's per capita emissions have been low, and Japan has also been very active in utilizing the flexible mechanisms of the Kyoto Protocol. More recently, however, emissions have been rising because of the Fukushima nuclear power plant disaster in 2011. When all nuclear reactors were closed down, the use of fossil fuels increased greatly. Russia's CO_2 emissions declined by more than one-third during the difficult restructuring of its economy and political system in the 1990s but then increased by about 7 per cent from 2000 to 2012. Russia's economy remains heavily dependent on fossil fuels, both for domestic use and export income. Russia is rich in renewable energy resources, but they are largely untapped.

Second, these seven actors have taken on crucial roles in the negotiations throughout the history of climate change diplomacy, albeit in different ways. Since the late 1980s, the European Union and the United States have gradually taken different roles, with the European Union perceived as an ambitious pusher while the United States has been perceived as a laggard. The most contentious issue in the process leading up to the adoption of the UNFCCC in 1992 was whether to establish binding emission targets. The United States was strongly opposed, arguing instead for a bottom-up approach. Japan proposed a bottom-up pledge-and-review system as a compromise between the top-down and the bottom-up approaches. At that time, however, the green NGO community and the European Union scorned this proposal (Bodansky, 1993). The compromise reached was a soft political pledge to aim for a stabilization of developed countries' emissions by 2000, much in line with the US position. In the process leading up to the adoption of the 1997 Kyoto Protocol, the European Union was fairly successful in pushing for ambitious targets, allowing developing countries, including emerging economies like China, India and Brazil, exemption for commitments, while the United States was the key architect behind the new flexible mechanisms. A detailed study of the process concluded that the United States was the most influential actor in designing the Protocol (Grubb et al., 1999).

In 2001 the United States rejected the Protocol on the grounds that it was too expensive to implement and unfair because it exempted major developing countries from commitments. With the United States out of the Kyoto process, a de facto new 'alliance' emerged between the United States and the G-77, including India and China, with the common objective not to accept any commitments. Still, the Kyoto Protocol came into force in 2005 when Russia ratified it. When negotiations on a second commitment period for the Kyoto Protocol started, it was once again the European Union that took a leadership role. The United States and the G-77 continued to refuse legally binding commitments to cut their emissions. The South also lashed out against the rich countries for not living up to their commitments to reduce emissions and assist them with financial and technological aid. No doubt, the European Union had leadership ambitions, but was not able to generate followers, a key criterion to qualify as leader (Underdal, 1994).

At COP 15 in Copenhagen, the cleavage among key actors was evidenced by tense discussions and a near breakdown of negotiations. Ultimately, an accord drafted by heads of state in the very last minutes of the summit was accepted (Christoff, 2010). The final deal was hammered out behind closed doors between a handful of countries: the United States and the so-called BASIC countries (Brazil, South Africa, India and China). In the Copenhagen Accord, the focus shifted away from a top-down–oriented way of collaboration towards more of a bottom-up approach that permitted countries to pledge how much emissions reductions they perceived as politically feasible by 2020, allowing a review process to ensure compliance (Olmstead and Stavins, 2012). Ironically, the US position of the late 1980s and the Japanese proposal of 1991 have re-emerged to serve as the main guidelines for the climate change negotiations 25 years later.

Third, and related to the two previous reasons, these seven actors are very likely to attract followers in the negotiations. Because of their size and significant positions, these actors are capable of forming a pivotal group that could persuade or induce other countries to join them in a new climate change agreement. Thus, we may safely assume that any agreement signed and ratified by all seven actors would be accepted by a large majority of other states as well. For instance, a US decision to sign and ratify would likely persuade their traditional followers in the negotiations, like Australia and Canada, to do the same. EU ratification would probably also induce Kyoto 2 partners to follow suit. Similarly, if China and India accept the terms in a new treaty, most if not all of the other G-77 members would likely follow their example.

CONCEPTUAL FRAMEWORK

In some instances, a quick look at countries' mitigation and damage costs will be sufficient to understand why countries' climate change policies often diverge. It can be no surprise that countries highly vulnerable to negative consequences of climate change (e.g., low-lying islands such as the Maldives) work very hard to put an effective global mitigation programme in place immediately while countries whose wealth is based mainly on the petroleum sector (e.g., Saudi Arabia) see such a programme as a severe threat. In addition, when comparing climate change policies of countries that are similar in most respects (e.g., the United Kingdom and the United States), their estimates of own mitigation and damage costs will likely provide important clues. In general, we may expect all governments, industries and other important stakeholders to care about costs and benefits and to include such estimates in their assessments of alternative climate policy options. Therefore, research aimed at explaining why some countries are willing and able to do more than others cannot afford to ignore these factors.[1]

Yet, models conceptualizing countries as unitary maximizers of own material welfare have important limitations. First, within all countries, mitigation and damage costs will be unevenly distributed. Moreover, the distribution of mitigation costs will in some cases vary significantly across policy options. When costs are unevenly distributed, internal policy preferences will likely diverge sufficiently to render the notion of a uniform national interest a crude and often misleading abstraction.

Second, even though policy preferences tend to reflect material self-interest, most actors are to some extent guided also by other concerns, notably values and norms (Dannenberg et al., 2010). For example, in the global climate change negotiations, certain norms of distributive fairness – including the polluter-pays principle and the ability-to-pay principle – are frequently invoked and rarely disputed (Lange et al., 2007; Mattoo and Subramanian, 2012). In a world characterized by stark asymmetries between rich (polluters) and poor (victims) the latter cannot be expected to accept proposals that do not reflect these and other broadly accepted equity norms. Furthermore, surveys show that public beliefs about likely impacts of climate change for future human generations vary significantly. In Pew's 2013 Global Attitudes survey, 85 per cent of the South Korean respondents and 76 per cent of the Brazilians listed global climate change as one of their most serious concerns, while the corresponding figures were 40 per cent in the United States and 39 per cent in China. The overall pattern is only in part consistent with scientific estimates of vulnerability.

Third, in decision-making processes, preferences, beliefs and positions are weighted by the relative power and influence of the actors involved. In all countries – including those ranked as the most advanced democracies – some institutions and parties have more political power and influence than others do. Further, since mitigation and damage costs are unevenly distributed, we may infer that costs and benefits that are concentrated to important sectors of the economy and/or to powerful segments of society tend to weigh more heavily than do costs and benefits concentrated to less important sectors/segments or spread more randomly throughout society (see Wilson, 1973; Evans, 1993).

Finally, political systems differ with regard to the concentration of power, along the vertical as well as the horizontal axis. Some countries grant more autonomy to regional and local authorities than others. The familiar distinction between federal and unitary states captures two main levels of centralization. Similarly, some countries (democracies) choose their political leaders through free elections and have established more or less complex systems of governance checks and balances while other countries (autocracies) are ruled by small and more or less self-recruiting elites. These institutional properties are of interest here primarily to the extent that they affect the political feasibility of policy change. They can do so through general rules specifying how leaders are (s)elected, what is required to block a certain policy option, and what is required to have an option formally adopted and effectively implemented. Other things being equal, more political energy will normally be required to adopt and implement a new policy than to continue on the existing path. The less open the (s)election of leaders, the larger the number of veto players that have to be satisfied, and the more radical the proposal, the larger this gap is likely to be. The latest assessment reports from the IPCC (2013) conclude that major cuts in GHG emissions are urgently required to prevent what the UNFCCC labels 'dangerous anthropogenic interference with the climate system'. Other things being equal, countries with high institutional barriers to radical change in established policies and constitutional arrangements will be less likely than countries with lower hurdles to respond effectively to IPCC's advice.

Politics Matters

The observations made in the preceding paragraphs suggest that countries' policies and practices may well deviate – under some circumstances significantly – from what unitary rational actor models would lead us to expect. To understand why and when such deviations occur we have to study how policies are made and implemented (Allison, 1969). States are, albeit to varying degrees, complex systems where actors sometimes collaborate and sometimes compete to influence decisions important to them. The policies emerging from these processes can best be understood as weighted aggregates of participants' preferences and beliefs, and these preferences and beliefs may reflect private as well as national and international concerns. System properties screen and weigh domestic concerns and preferences in ways that may lead countries with similar scores on the national mitigation and damage costs variables to pursue different climate change policies (see Harrison and Sundstrom, 2010, p. 3). To identify and understand the mechanisms at work here we have to open the black box of unitary rational actor models and engage in comparative analysis of domestic politics.

In doing so, we should take care not to discard the fundamentally important insight expressed in the mitigation/damage costs formula. We have therefore designed this study so as to systematically assess how domestic institutions and processes shape mitigation policies *in the context of* important material parameters, in particular countries' energy resource endowments.

Interests, Institutions and Ideas

The field of comparative politics research provides a rich reservoir of analytical tools and substantive propositions and findings that can facilitate and guide our analysis. In the sphere of analytical tools, important contributions range from specific concepts (such as 'veto player' and 'winning coalition') to applications of overarching theoretical frameworks (such as 'rational-choice analysis' and 'constructivism'). Following Steinberg and VanDeveer (2012), we argue that a common conceptual framework can provide important tools for describing similarities and differences in terms of scores on the same variables and thereby help researchers tap into reservoirs of knowledge and insight accumulated through previous studies of other cases. This applies, we submit, even when the task is one of describing and understanding policy-making in countries whose political, economic and cultural systems differ widely.

Developing a conceptual framework that can serve this function requires substantive knowledge that identifies critical variables and the causal mechanisms through which they work. Substantive contributions range from very general models identifying principal clusters of independent variables to analytical propositions and empirical findings about one particular pattern of variance. At the most general level, comparative politics research focuses much of its efforts on understanding the roles of 'rationality, culture, and structure' (Lichbach and Zuckerman, 2009). In two recent contributions to comparative analysis of environmental governance, these clusters are referred to as 'interests, ideas, and institutions' (Harrison and Sundstrom, 2010; Steinberg and VanDeveer, 2012). As indicated by the labels used by Lichbach and Zuckerman, the three clusters are often linked to competing 'research paradigms', most clearly in the cases of 'rationality' (rational-choice analysis) and 'culture' (constructivism). Such links can be found also in the field of environmental governance but here a more ecumenical and pragmatic approach seems to prevail. Interests, ideas and institutions call for attention to the extent that they are important determinants of policies and practices and should be studied by whatever tools lead to the most interesting and robust insights. Heeding the advice of Fearon and Wendt (2002), we adopt this ecumenical and pragmatic approach.

At the level of specific findings, comparative politics research offers conclusions derived through formal modelling as well as findings based on empirical evidence. Riker's (1962) 'minimal winning coalition' proposition and Tsebelis's (2002) analysis of veto players are well-known examples of the former. Similarly, Lachapelle and Paterson's (2013, p. 563) finding that 'parliamentary systems were more likely than presidential ones to implement a range of climate policies to mitigate emissions' and Madden's (2014) finding (in support of Tsebelis's argument) that 'multiple political institutions tend to lead to lower climate policy adoption rates' may serve as examples of the latter.

Our case studies make extensive reference to all these categories of contributions, as does our search for more general patterns in the concluding chapter. The conceptual framework outlined below has, however, been adapted and designed specifically for the purposes of this particular study. The book explores prospects of policy *change*, more specifically the development of mid-term policies that drive or block transition towards low-carbon energy systems. In particular, we seek to understand the role of climate change concern as a driving force for policies aimed at energy system change. Moreover, our framework is designed to take into account *distinctive* features of the climate change mitigation challenge,[2] in particular the fact that the anthropogenic impact on the climate system can be traced to a very wide range of human activities – from burning of fossil fuels to land use and land cover change. Consequently, mitigation policies are inextricably linked to many other policy domains. To succeed, mitigation policies must somehow penetrate these domains and their established networks. Combining the latter observation with our interest in drivers for and barriers to policy change has three important implications for the kind of conceptual framework needed.

First, we need a framework that can help identify stable or slow-changing parameters that constitute important boundary conditions for policy change. Here we examine two main categories of such boundary conditions. One includes material parameters such as a country's (energy) resource endowments, material infrastructure and wealth. The other includes political settings, notably the institutional structure and the distribution of power. Both kinds of parameters will likely serve as short- to medium-term stabilizers of each state's policies and account for a significant proportion of the variance found when we compare states' policies.

Second, we need a framework that can help identify the principal forces that drive or block policy change and explain how they work. We propose that in all political systems, policy development can be understood as driven by two main generic forces: (societal) demand and (governmental) supply (Underdal and Hanf, 2000). These forces interact and co-produce policies, but the ways they do so vary significantly, depending on the characteristics of political institutions, social structures, culture and other country-specific features (see Chapter 9).

Finally, to study policy trajectories and conditions for change, we need a framework that can help us understand political processes as evolving sequences of connected events in which early decisions and actions can affect the menu of choice for subsequent decisions and actions (Pierson, 2000). For example, accumulated investments in material infrastructure – like power plants, grids and transportation systems – can sometimes lock countries or industries to a particular path.

Fundamental Parameters

As defined above, a parameter is a stable or slow-changing factor that constitutes an important boundary condition for policy decisions. Material parameters are fundamental in that they are important determinants of costs and benefits of alternative policy options. Political parameters are less stable than their material counterparts are, and instances of revolutionary change do occur.[3] Yet, for most countries, radical institutional reforms occur infrequently and so do abrupt and profound changes in the distribution of domestic political power.

For climate change mitigation policy, material interests can be conceived as a function of the expected damage caused by climate change and the relative costs of mitigation and adaptation measures. A rational actor motivated by self-interest will pursue mitigation as long as, and only as long as, its marginal mitigation costs are lower than its marginal damage and adaptation costs. Yet, the story is often complicated. First, actors will likely consider other types of costs and benefits. In some models, particularly those inspired by economics, policy-makers are assumed to be concerned with maximizing votes in elections or with maximizing support and status more broadly defined (Downs, 1957). Second, policy decisions are premised on beliefs and may be guided by values and norms too. Thus, to understand why gasoline taxes that would appear insignificant to most Scandinavians create widespread, vocal resistance in the United States we should look not merely at monetary costs but also at values and beliefs embedded in culture and lifestyles. More generally, willingness to pay for mitigation measures depends on (1) the value attributed to certain environmental assets and services and (2) one's beliefs about the role of human activities in causing environmental damage, one's vulnerability to likely environmental changes, and one's relative costs of mitigation and adaptation.

Principal Forces

Studying policy development as driven by demand and supply involves determining the direction, strength and sources of each of these forces and analysing the interplay between them.

An actor's *demand* for a certain policy or action is an expression of its preference for that policy or action over other available options.[4] *Aggregate* demand is the sum total of all actors' demands for a particular option, weighted for each actor's relative power, and the capacity for collective action that groups of actors have. *Relative* demand is the strength of the aggregate demand for one policy option over another.

In extant research, demand is most often seen as emanating from societal sources and policy supply as being in governments' hands. Here, we think of both in more generic terms, allowing for demand to come also from governmental sources and for non-governmental parties to be actively engaged on the supply side. Political controversies typically engage more or less domain-specific coalitions including governmental as well as non-governmental actors (Sabatier, 1988). Moreover, countries' policies and practices to some extent co-evolve, through unilateral learning and adaptation as well as through cooperation.

Democracies provide their citizens with a wider range of opportunities for openly expressing demands – particularly demands for change – than authoritarian systems do. Democracies also provide their citizens with more effective sanctions on election day should the government fail to deliver. Consequently, societal demand will usually be a much weaker driver of policy development in authoritarian systems than in democracies. This difference is, though, a matter of degree. All political systems have a 'selectorate' choosing state leaders, and even autocrats and ruling party elites have good reasons to worry about widespread public dissatisfaction and emerging cracks in their own power base (Bueno de Mesquita et al., 2005). In addition, latent demand may be politically significant, particularly if found within the regime's principal constituency. Societal demand need not be vocally articulated to make a difference.

(Governmental) supply may be conceptualized similarly. The climate and energy policies pursued by a government will depend on the preferences and beliefs of governmental actors (branches, coalition partners, etc.), the distribution of authority and power among these actors, and the extent to which the government controls the systems of activities to be governed. Authoritarian systems will typically concentrate more power in the hands of a small ruling elite (in extreme cases, one ruler) than democracies do, and also provide this elite with far more effective means for controlling industrial and societal activities than are available in democracies. By implication, democratic systems tend to be more susceptible than their authoritarian counterparts are to 'vertical disintegration' of policies, that is, to a state of affairs where the aggregate thrust of micro-decisions deviates more or less significantly from what policy doctrines and principles would lead us to expect (Wilson, 1973; Underdal, 2000, pp. 72–4). Moreover, while democratic systems assign distinct and different roles to key institutions – notably the executive, the legislative and the judiciary – similar separation of powers formulas are at best muted in authoritarian and totalitarian systems. A conceptual framework must be sufficiently flexible to accommodate such differences.

Demand and supply co-produce policies through a system of political institutions that lays down decision-making rules. The main functions of an institution are to define a particular practice (such as policy-making), to assign roles to participants involved in this practice (such as cabinet ministers and members of parliament), and to guide interactions among occupants of these roles through decision rules and procedural rules (Young, 2002, p. 5). In brief, an institutional system of governance specifies *who* is authorized to make decisions about *what*, and *how* these decisions shall be made. The relative importance of demand-side and supply-side mechanisms in shaping policies and practices differs significantly. Yet, even in non-democratic systems, where the balance is heavily skewed in favour of supply-side mechanisms, governmental supply will not be immune to demand from the regime's core constituencies.

To grasp the essence of demand–supply interplay, we combine Sabatier's analysis of advocacy coalitions with Tsebelis's analysis of veto players (Sabatier, 1988, 2007; Tsebelis, 2002). Given our interest in the conditions for policy change, two types of actors become particularly important: veto players and winning coalitions. Tsebelis defines 'veto players' as actors whose agreement is necessary for a change of the status quo (Tsebelis, 2002, p. 19). Building on his analysis, we conceive of the stability of an established policy as a function of (1) the number of veto players, (2) the policy distance(s) between them[5] and (3) their collective action capacity. To determine the feasibility of a particular policy change option, it makes sense to begin by determining whether the status quo is preferred over the alternative policy by one or more veto players. One veto player is sufficient to block a particular proposal. A proposal may, however, be amended and/or linked to other policy issues to satisfy this pivotal actor. To evaluate conditions and prospects for change, we will often need a more comprehensive analysis of the political landscape and policy-making processes.

The inverse of Tsebelis's formula can be used to explore the scope of feasible policy change, although only as a first cut. To be adopted and implemented, a policy change option need not merely escape all veto players; it must also attract the active support of a 'winning coalition'. Since the status quo is the default option, it can prevail even if no veto player protects it; in the absence of collective action capacity, a 'silent' majority may well just remain silent. Here, Sabatier's analysis of advocacy coalitions offers useful guidance. Sabatier defines advocacy coalitions as being 'composed of people from various organizations who share a set of normative and causal beliefs and who often act in concert' (Sabatier, 1988, p. 133). Members of an advocacy coalition 'will show substantial consensus on issues pertaining to the policy core' but less so on issues pertaining to peripheral elements (ibid., p. 146). Sabatier identified two main conditions for 'significant' change of 'the core' of an established policy programme: (1) a shift of power from the coalition that instituted the programme to a rival coalition; (2) significant changes in socioeconomic conditions or new external policy inputs. The institutional setting determines what is required to 'win'.

In sum, the discussion above provides us with a set of assumptions regarding factors that are expected to influence the potential for policy change in the seven key actors. We expect material parameters like domestic energy resource endowments and related concerns about domestic energy production, distribution and consumption to be an important cluster of explanatory factors for climate policy trajectories. Energy needs in the electricity sector seem to play a particularly important role for policy-makers, and we see material interests as a function both of the expected damage caused by climate change and the relative cost of mitigation and adaptation measures. In other words, we expect the level of mitigation costs to affect decisions, especially in the energy sector. The more expensive policy measures are, the more they negatively affect energy prices, and the more they hinder the use of domestic energy resources, the less likely they are to be enacted. We expect that this effect is also linked to the degree of fossil fuel dependency each actor experiences, since fossil fuels traditionally are cheaper energy sources than low-carbon or carbon-free energy sources.

Since political systems are different with respect to the degree of fragmentation along both the horizontal and vertical axes, we assume that the degree of fragmentation will affect the likelihood of climate policy change. Specifically the degree of unity in decision-making power influences the policy-making process, where more centralization of powers will facilitate policy change. Conversely, political systems with a high degree of multilevel governance have more veto points in the process. There are also crucial differences between non-democracies and democracies. Democracies have a much wider range of channels and opportunities for societal actors to express demand and support. More societal actors are involved in political processes, and policy-makers have strong incentives to pay attention to their voters' concerns.

Concerning societal demand, the assumption is that the strength and form of societal demand influence the ambitiousness and likelihood of climate policy change. In democracies, societal demand is more often clearly articulated by a wide range of societal actors, and we assume that policy change will be more likely in countries where advocacy coalitions are stronger than veto players are. However, we also presume that even autocrats and ruling party elites must be concerned about social unrest and about avoiding public dissatisfaction that can threaten their power base. Hence, societal demand plays out differently in autocracies, where public opposition is rare, and where involvement of societal actors happens in other ways.

Finally, we expect that active supply of new governmental initiatives will increase the likelihood of climate policy change. Policies pursued by government depend on the preferences and beliefs of governmental actors, the distribution of authority and power among these actors, and the extent to which government controls the activities to be regulated. In general, we expect that if many governmental actors are empowered or have jurisdiction over a topic, the process will be more complicated than if only a small number of actors have power to propose new policies. We expect there to be clear differences between democracies and autocracies in this respect, with clear separation of powers in democracies and no real power distribution between governmental branches in autocracies. Moreover, we expect preferences to be different in developed countries from those in developing countries that have an urgent priority for economic growth.

The 'Null Hypothesis'

Clearly, we take as an important underlying assumption that government policies play distinct and significant roles in shaping countries' GHG emission practices and trajectories, even if in part reflecting fundamental material parameters. Although intuitively plausible, a critical look at that assumption seems warranted.

One of the most important challenges to our 'policy matters' assumption can be found in (interdisciplinary) research adopting a more holistic Earth system perspective. More than four decades ago researchers studying longer-term (macro) patterns developed a proposition suggesting that the aggregate human impact on the environment (I) is determined primarily by population size (P), affluence (A, usually measured as GDP per capita), and technology (T) (see e.g., Commoner et al., 1971; Holdren and Ehrlich, 1974; Chertow, 2001). These three factors are supposed to interact as expressed in the formula $I = P \times A \times T$. Applied to the climate change domain, the reasoning behind this formula may be summarized as follows:

1. Other things being equal, GHG emissions tend to increase with population size and affluence. The increase with population size will likely be monotonous but not necessarily linear. The increase with affluence will likely be non-linear and may follow the pattern known as the environmental Kuznets curve (EKC). The EKC hypothesis postulates an inverse U-shaped relationship between (national) income and pollution, implying that pollution tends to rise (steeply) as affluence increases from a very low level but then

tends to gradually flatten and eventually decline in rich societies (Kuznets, 1955).[6]

2. The impact of population size increases with affluence, and vice versa. Their *joint* effect is therefore best understood as multiplicative rather than additive.
3. Technology can work in both directions, to reduce as well as to increase emissions. For example, technological development can increase energy efficiency and thus reduce GHG emissions per unit of production but it can also enable large-scale tapping of natural resource deposits that were previously unknown or not considered attractive for industrial purposes, and enhance the productivity of labour and thereby raise income levels. In applying the I = PAT formula to a specific environmental domain, further specification of the assumed role(s) of technology in that particular setting will be required. Overall, though, technology is seen as serving an enabling function that can best be modelled as one of weighting the environmental impact of population and affluence.

Critics of the I = PAT formula have pointed out that it fails to recognize important aspects, in particular interdependence relationships existing between the three components and – partly as a consequence – also potential rebound effects (see, e.g., Waggoner and Ausubel, 2002; Alcott, 2010). Some researchers have tried to further develop the formula by adapting it to specific characteristics of a particular environmental domain. For climate change, one friendly amendment includes, as a fourth determinant of human impact, CO_2 emissions per unit of energy consumed (see, e.g., Kaya, 1990; IPCC, 2001). For our study it is important to note that all the driving forces included in the formula are to some extent influenced by government policies. In this sense, the formula does not provide a completely policy-free alternative. On the other hand, the formula itself has nothing to say about policies and politics, so for our study it offers no real alternative to the framework outlined above. Nevertheless, by focusing on essential features of the development of humankind, the formula does provide a more holistic model for understanding Earth system dynamics in an era often referred to as the 'Anthropocene'. Since the formula is also sufficiently well-grounded in natural as well as social science theory to merit serious attention, we will treat it as a 'null hypothesis', to be examined in the concluding chapter.

OVERVIEW OF THE BOOK

The book's individual contributions are organized into eight further chapters covering case study analysis of each of the seven key actors.

In Chapter 2, Solveig Aamodt assesses the policy trajectory of Brazil. Brazil adopted a comprehensive climate law in 2009, one of the first countries to do so. With specific mitigation targets, the law followed up successful implementation of policies to reduce deforestation. Emissions from deforestation decreased by 75 per cent between 2004 and 2012, reducing Brazil's aggregate GHG emissions significantly. However, while deforestation has declined, GHG emissions from agriculture and energy increased over the past decade and this growth is expected to increase at a faster rate after 2020, especially in the energy sector. Aamodt analyses drivers of and barriers to mitigation policy in different emission sectors in Brazil. Several low-carbon solutions are more profitable in Brazil than in many other countries, but concerns regarding energy security and food security constrain the prioritization of low-carbon solutions. Brazilian politicians feel little pressure to increase mitigation targets beyond what is achieved through reduced deforestation, and well-established interest groups in society prefer traditional solutions to low-carbon innovation. Current policy decisions have a large impact on Brazil's post-2020 emission trajectories and Aamodt assesses how the current political situation influences future mitigation possibilities.

In Chapter 3, Iselin Stensdal discusses recent developments in China's climate and energy policies. China is the world's largest CO_2 emitter, because of unparalleled economic development since the 1980s. GHG mitigation and energy policies are becoming more ambitious. The main low-carbon measures are reshuffling the economy and energy mix, as well as improving China's energy and carbon intensities. The mitigation policies are mainly top-down driven: the central government supplies local governments with measures to execute. Beneath the central government level, ministries and local governments advocate for their own interests, at the same time as being required to emphasize energy saving and emissions reduction as national priorities. Researchers at select institutions and environmental non-governmental organizations (ENGOs) display supportive demand for mitigation policies. Recently the public's awareness of and dissatisfaction with air pollution and PM2.5 (fine particle pollution), even expressed online via Weibo, has spurred a swift supply of air pollution mitigation policies. Most pollution reduction measures will also reduce GHG emissions; mitigation occurs as a co-benefit of air pollution mitigation. For the central government the challenge is to make economic development and environmental protection feasibly compatible.

In Chapter 4 the European Union and its climate and energy policy package are studied. Jon Birger Skjærseth analyses the European Union's leadership-by-example role in international climate negotiations and how the European Union has managed to have a relatively ambitious climate policy accepted by 28 member states with widely differing energy economic situations. Energy security concerns spur a need for energy efficiency and higher energy production, but member state views differ regarding how shifting to a low-carbon economy will promote energy security. The institutional setting provides an enabling context, stimulating consensus seeking and long-term policy development. Because EU policies were adopted by consensus from 2007, explanations for change focus on how climate and energy policies and issues were combined in new ways that enabled cost sharing, promoted new low-carbon opportunities and gave something to all major veto players. The linkage between climate and energy policies has mainly been policy supply–driven, but broad support for EU-level climate policies has been important for legitimizing the decisions taken. The European Union's new 2030 climate and energy policy framework represents a policy 're-packing' compromise to satisfy the main veto players, with substantial concessions to Poland and other Central and Eastern European Countries (CEECs). Whether new policies can put the member states collectively on the path towards a low-emission economy, however, will depend on new legislation yet to be adopted and implemented.

In Chapter 5, India's climate and energy policy is analysed by Sunil Tankha and Trude Rauken. India is the world's largest democracy with more than 1.2 billion inhabitants, but with one-third of the population lacking access to modern energy services. India's development goals and needs will trump climate change issues, meaning that India is unwilling to make commitments to reduce emissions under the UNFCCC. India argues for its right to development, but at the same time India is highly vulnerable to climate change impacts, putting climate change on the agenda of Indian decision-makers. In being the world's third-largest emitter, India plays an integral part in future emission scenarios. Still, there are few signs of large emission cuts in India. India has a fossil fuel–dependent energy production and there is little political leeway to cut emissions if doing so is seen as limiting economic growth. Instead, India's policy-makers tend to focus more on adaptation, and civil society and the general citizenry in India are not in favour of emissions curbs for mitigation purposes. Thus, international pressure alone is not likely to lead to a strong commitment to cut emissions in India. Instead, India will keep economic growth as its primary goal, but if emission-reducing measures are seen as economically profitable, India is more likely to embrace them.

Japan's climate and energy policy is analysed in Chapter 6, where Masahiko Iguchi, Alexandru Luta and Steinar Andresen assess drivers of and barriers to policy action in the wake of the Fukushima accident. Japan's climate policy has experienced significant changes. Frequent shifts of governments helped bring about various emission reduction pledges. However, only the short-lived Democratic Party of Japan (DPJ) government had a truly ambitious policy, although the most ambitious parts were never implemented. A major reason was the 2011 Fukushima Daiichi Nuclear Disaster, which altered the priorities of Japanese politics. As a result the Liberal Democratic Party (LDP) in 2013 proposed a very modest target, compared to those of other states as well as to previous pledges. It should be borne in mind that the costs of reducing emissions in Japan are particularly high because of its high energy efficiency score. Japan has relied primarily on voluntary policies, some of which have been quite effective. Still, the actors demanding a more forceful climate policy are weak and dispersed, whereas the supply-side veto players are strong and concentrated. There are few indications that this situation is about to change in the near future. Renewable energy cannot expand quickly enough to make up for the shortfall from nuclear power. Thus, if emissions are to be reduced, nuclear power will probably play a more important role than will renewable energy.

Anna Korppoo analyses Russian climate and energy policy in Chapter 7. The prospects for significant policy change towards decarbonization are low in Russia due to the lack of environmental concern, in particular over the impacts of climate change, the interpretation of the post-Soviet emission decline as a climate effort, and the dominance of the economically vital fossil fuel sector. The main emission reductions are likely to originate from cutting energy waste and from modernization rather than focused mitigation policies. However, the abundance of domestic fossil fuels, the unfavourable investment climate and the weakness of the policy implementation system obstruct these policies. Nevertheless, Russia's domestic goal to limit GHG emissions to 75 per cent of 1990 level by 2020, which never deviated much from the business-as-usual emissions trend, is likely to be achieved more easily than expected due to the current negative economic prospects. Even though the political conflict with Ukraine has destabilized President Putin's power, and an overthrow of the Putin regime seems less unrealistic than before the conflict, climate policy is unlikely to change due to the lack of public demand for stronger measures. At the international level Russia is likely to wait and see how other major emitters will play their cards in Paris in 2015 before deciding on its participation in a new climate agreement.

In Chapter 8, the final country study in this book is devoted to the United States, which has experienced several attempts at climate and energy policy action under the Obama administration. Guri Bang explores the domestic politics that underpins US positions in international climate negotiations, and the drivers for and barriers to realizing the Obama administration's proposed climate policy changes. The United States is the world's second-largest emitter of GHGs, with ample domestic coal, shale gas and shale oil reserves. Increased use of natural gas to replace coal in the power sector resulted in declining levels of energy-related CO_2 emissions from 2009 onwards. Deep reliance on fossil fuel energy, in combination with intense polarization between Republicans and Democrats on climate policy issues, strong opposition to climate action from key stakeholder groups, and lack of public pressure to act on the problem are the most important barriers to an ambitious domestic climate policy in the United States. Despite these profound barriers, the Obama administration proposed significant changes in the approach to climate change policy. Obama used his executive powers to develop new regulations aimed at cutting CO_2 emissions through existing law. As a result, state-specific CO_2 regulations for power plants are being developed for the first time in the United States. While public pressure for more federal climate action has slowly increased in step with the recovery of the US economy after the financial crisis, deep disagreements that have dominated US climate policy deliberations among federal law-makers over the past 20 years persist.

In Chapter 9, Guri Bang, Arild Underdal and Steinar Andresen revisit the assumptions made in this introductory chapter to compare and contrast similarities and differences between the key actors on the basis of the findings in the country studies. First, they show that the I = PAT formula identifies and combines important sources of environmental change and can help predict and understand main emission trajectory patterns at the macro level. Predictions of aggregate world GHG emissions show good fit with observed trajectories. At the level of individual countries, however, results are far more diverse, and even show some recent instances of decoupling of emission trajectories from the formula. Second, therefore, the authors assess why the conceptual framework applied in the case studies offers better explanations, highlighting domestic politics and government policies.

Analysing the interplay between fundamental parameters and driving forces, the chapters of this volume provide a comprehensive picture of the drivers and barriers in domestic climate policy and the prospects for change in the short to medium term. By first defining basic parameters that set the scope for energy use and policy interaction, and then adding layers of more complex information about the mechanisms and conditions for governmental supply and societal demand, policy trajectories and the prospects for policy change are assessed.

NOTES

* We thank Jon Hovi for excellent and helpful comments to an earlier version of this chapter.

1. This is not to say that they are to be weighted equally. Except for the most vulnerable actors (short-term) mitigation costs seem to attract more attention than do (long-term) damage costs.
2. Steinberg and VanDeveer (2012, p. 30) describe general comparative politics research and comparative environmental politics research as 'separate archipelagos'. One indication is the scant attention paid to the environmental politics domain in main journals. In *Comparative Politics* and *Comparative Political Studies*, 1 per cent and 2 per cent respectively of all articles published 1990–2010 focused on environmental matters, while 36 per cent and 34 per cent, respectively, dealt with 'economic policy and growth' (Steinberg and VanDeveer, 2012, pp. 373f).
3. However, as the rapid development of the shale gas industry illustrates, estimates of technologically feasible and economically attractive energy resource endowments may change quite rapidly.
4. This definition includes support for current policy.
5. In other words, the more veto players there are, and the larger the policy distances between (among) them, the larger becomes the range of policy change options that at least one of them can effectively veto.
6. Empirical tests of the EKC hypothesis have yielded somewhat divergent results, overall showing more support in studies of local pollution than in studies dealing with global commons (see e.g., Dasgupta et al., 2002; Harbaugh et al., 2002; Stern, 2004; Carson, 2010).

REFERENCES

Alcott, B. (2010), 'Impact caps: why population, affluence and technology strategies should be abandoned', *Journal of Cleaner Production*, **18**(6), 552–60.

Allison, G. (1969), 'Conceptual models and the Cuban Missile Crisis', *American Political Science Review*, **63**(3), 689–718.

Andresen, S. (2015), 'International climate negotiations: top-down, bottom-up or a combination of both?', *The International Spectator: Italian Journal of Foreign Affairs*, **50**(1), 15–30.

Bodansky, D. (1993), 'The United Nations Framework Convention on Climate Change: a commentary', *The Yale Journal of International Law*, **18**(2), 451–558.

Bueno de Mesquita, B., A. Smith, R.M. Siverson and J.D. Morrow (2005), *The Logic of Political Survival*, Cambridge, MA: MIT Press.

Carson, R.T. (2010), 'The environmental Kuznets curve: seeking empirical regularity and theoretical structure', *Review of Environmental Economics and Policy*, **4**(1), 5–23.

Chertow, M.R. (2001), 'The IPAT equation and its variants: changing views of technology and environmental impact', *Journal of Industrial Ecology*, **4**(4), 13–19.

Christoff, P. (2010), 'Cold climate in Copenhagen: China and the United States at COP 15', *Environmental Politics*, **19**(4), 637–56.

Commoner, B., M. Corr and P.J. Stamler (1971), *The Closing Circle: Nature, Man, and Technology*, New York: Knopf.

Dannenberg, A., B. Sturm and C. Vogt (2010), 'Do equity preferences matter for climate negotiators? An experimental investigation', *Environmental and Resource Economics*, **47**(1), 91–109.

Dasgupta, S., B. Laplante, H. Wang and D. Wheeler (2002), 'Confronting the environmental Kuznets curve', *Journal of Economic Perspectives*, **16**(1), 147–68.

Downs, A. (1957), 'An economic theory of political action in a democracy', *Journal of Political Economy*, **65**(2), 135–50.

Evans, P.B. (1993), 'Building an integrative approach to international and domestic politics: reflections and projections', in P.B. Evans, H.K. Jacobson and R.D. Putnam (eds), *Double-Edged Diplomacy: International Bargaining and Domestic Politics*, Berkeley and Los Angeles, CA: California University Press.

Falkner, R., S. Hannes and J. Vogler (2010), 'International climate policy after Copenhagen: towards a "building blocks" approach', *Global Policy*, **1**(3), 252–62.

Fearon, J. and A. Wendt (2002), 'Rationalism v. constructivism: a skeptical view', in W. Carlsnaes, T. Risse and B.A. Simmons (eds), *Handbook of International Relations*, London: Sage.

Grubb, M., C. Vrolijk and D. Brack (1999), *The Kyoto Protocol. A Guide and Assessment*, Amsterdam: Royal Institute of International Affairs.

Gupta, J. (2010), 'A history of international climate change policy', *WIREs Climate Change*, **1**(5), 636–53.

Harbaugh, W.T., A. Levinson and D.M. Wilson (2002), 'Re-examining the empirical evidence for an environmental Kuznets curve', *Review of Economics and Statistics*, **84**(3), 541–51.

Harrison, K. and L.M. Sundstrom (eds) (2010), *Global Commons, Domestic Decisions: The Comparative Politics of Climate Change*, Cambridge, MA: MIT Press.

Holdren, J.P. and P.R. Ehrlich (1974), 'Human population and the global environment', *American Scientist*, **62**(3), 282–92.

International Panel on Climate Change (IPCC) (2001), *Special Report on Emission Scenarios: Summary for Policymakers*, Geneva: IPCC, accessed 23 April 2015 at https://www.ipcc.ch/pdf/special-reports/spm/sres-en.pdf.

International Panel on Climate Change (IPCC) (2013), *Fifth Assessment Report (AR5)*, Geneva: IPCC, accessed 27 March 2015 at www.ipcc.ch/report/ar5/.

International Panel on Climate Change (IPCC) (2014), 'National and sub-national policies and constitutions', in *Climate Change 2014: Mitigation of*

Climate Change. Working Group III Contribution to the IPCC 5th Assessment Report, Geneva: IPCC, accessed 23 April 2015 at https://www.ipcc.ch/report/ar5/wg3/.

Kaya, Y. (1990), 'Impact of carbon dioxide emission control on GNP growth: interpretation of proposed scenarios', paper presented to the IPCC Energy and Industry Subgroup, Response Strategies Working Group, Paris.

Kuznets, S. (1955), 'Economic growth and income inequality', *American Economic Review*, **49**(1), 1–28.

Lachapelle, E. and M. Paterson (2013), 'Drivers of national climate policy', *Climate Policy*, **13**(5), 547–71.

Lange, A., C. Vogt and A. Ziegler (2007), 'On the importance of equity in international climate policy: an empirical analysis', *Energy Economics*, **29**(3), 545–62.

Lichbach, M.I. and A.S. Zuckerman (eds) (2009), *Comparative Politics: Rationality, Culture, and Structure*, 2nd edition, Cambridge, UK: Cambridge University Press.

Madden, N.J. (2014), 'Green means stop: veto players and their impact on climate-change policy outputs', *Environmental Policy*, **23**(4), 570–89.

Mattoo, A. and A. Subramanian (2012), 'Equity in climate change: an analytical review', *World Development*, **40**(6), 1083–97.

Olmstead, S.M. and R.N. Stavins (2012), 'Three key elements of a post-2012 international climate policy architecture', *Review of Environmental Economics and Policy*, **6**(1), 65–85.

Pierson, P. (2000), 'Path dependence, increasing returns, and the study of politics', *American Political Science Review*, **94**(2), 251–67.

Riker, W.H. (1962), *The Theory of Coalitions*, New Haven, CT: Yale University Press.

Sabatier, P.A. (1988), 'An advocacy coalition framework for policy change and the role of policy-oriented learning therein', *Policy Sciences*, **21**(2–3), 129–68.

Sabatier, P.A. (ed.) (2007), *Theories of the Policy Process*, Boulder, CO: Westview Press.

Steinberg, P.F. and S.D. VanDeveer (eds) (2012), *Comparative Environmental Politics: Theory, Practice, and Prospects*, Cambridge, MA: MIT Press.

Stern, D.I. (2004), 'The rise and fall of the environmental Kuznets curve', *World Development*, **32**(8), 1419–39.

Tsebelis, G. (2002), *Veto Players: How Political Institutions Work*, Princeton, NJ: Princeton University Press.

Underdal, A. (1994), 'Leadership theory. Rediscovering the arts of management', in I.W. Zartman (ed.), *International Multilateral Negotiations: Approaches to the Management of Complexity*, San Francisco, CA: Jossey-Bass, pp. 178–97.

Underdal, A. (2000), 'Conceptual framework: modelling supply of and demand for environmental regulation', in A. Underdal and K. Hanf (eds), *International Environmental Agreements and Domestic Politics: The Case of Acid Rain*, Aldershot, UK: Ashgate, pp. 49–86.

Underdal, A. and K. Hanf (eds) (2000), *International Environmental Agreements and Domestic Politics: The Case of Acid Rain*, Aldershot, UK: Ashgate.

United Nations Framework Convention on Climate Change (UNFCCC) (1992), accessed 23 April 2015 at http://unfccc.int/resource/docs/convkp/conveng.pdf.

US Energy Information Administration (EIA) (2013), *US Energy-Related Carbon Dioxide Emissions, 2012*, accessed 25 January 2015 at http://www.eia.gov/environment/emissions/carbon/archive/2012/index.cfm.
US Energy Information Administration (EIA) (2014), 'Country analysis: China', accessed 28 March 2015 at http://www.eia.gov/countries/cab.cfm?fips=CH.
Victor, D. (2011), *Global Warming Gridlock. Creating More Effective Strategies for Protecting the Planet*, Cambridge, UK: Cambridge University Press.
Waggoner, P.E. and J.H. Ausubel (2002), 'A framework for sustainability science: a renovated IPAT identity', *PNAS*, **99**(12), 7860–65.
Wilson, J.Q. (1973), *Political Organizations*, New York: Basic Books.
World Resources Institute (2015), 'CAIT database: total GHG emissions excluding land-use change and forestry – 2011', accessed 27 March 2015 at http://cait2.wri.org.
Young, O.R. (2002), *The Institutional Dimensions of Environmental Change: Fit, Interplay, and Scale*, Cambridge, MA: MIT Press.

2. To be – or not to be – a low-carbon economy: a decade of climate politics in Brazil

Solveig Aamodt

BOX 2.1 KEY FACTS ABOUT BRAZIL

Population 2013	200.4 million
Population growth rate 2013	0.9%
Population density 2013	23.5 per km^2
GDP per capita 2013	11 208 USD
GDP growth rate	2.7% (annual average 2009–13)
GINI index 2009	54.7
Access to electricity 2011	99.3% of the population
Land area	8 459 420 km^2
Political system	Federal democratic republic
Subnational entities	26 states and the Federal District with the capital Brasilia

Source: World Bank (2014).

INTRODUCTION

Since the Rio Summit in 1992 'sustainable development' has been a buzzword in Brazilian politics, and the Brazilian population is among the world's most concerned about climate change (Ray, 2009). Brazilian policy-makers and important stakeholders rhetorically agree that Brazil's economy should be a low-carbon economy. They disagree, however, on how far the country has come on the low-carbon path.

In numbers (see Box 2.1 and Table 2.1) Brazil has mitigated more emissions than most other countries have; the aggregate emissions in 2013 are 33 per cent lower than the emissions in 2005. Emissions from the main source, land use change and deforestation, were 73 per cent lower in 2013 compared to the peak year 2004. However, emissions from agriculture, and particularly from energy, are growing and these sectors are each now responsible for around one-third of Brazil's aggregate emissions (SEEG, 2014).

Reduced deforestation emissions have put Brazil in an advantageous position internationally; as long as Brazil keeps deforestation stable, the target to reduce greenhouse gas (GHG) emissions by 36.1–38.9 per cent below business-as-usual (BAU) trajectories by 2020 is well within reach, even if energy emissions increase.

The mitigation debate is nevertheless ongoing domestically. The main line of conflict in Brazilian climate politics is the classical cleavage between growth-oriented interests and conservation-oriented interests, in combination with sector-related cleavages. With its basis in biodiversity protection and indigenous peoples' rights, the societal mobilization against deforestation and forest degradation has been strong, and in constant conflict with the interests of landowners and agribusiness.

Early investments in hydropower and biofuels have given Brazil a relatively clean energy matrix, and the energy sector has not been targeted for mitigation. Energy security and energy prices are the main concerns in Brazilian energy policy development. Increased use of fossil fuels in transport and electricity generation is altering the clean image, but the growth-oriented interests are still dominating energy policy development.

In emission projections for 2030, deforestation emissions are expected to stabilize at current levels or below, while agriculture and energy emissions are expected to increase, with incremental energy emission growth after 2020 unless new mitigation measures are taken (La Rovere et al., 2013). Current policy decisions greatly impact Brazil's post-2020 emission trajectories. This chapter analyses current drivers of and barriers to mitigation policy and low-carbon solutions in Brazil, presenting likely policy trajectories for 2020. Since energy emissions are expected to triple without mitigation, the majority of the analysis concentrates on the energy sector.

In this chapter, the next section outlines current climate policy and recent trends. The third section outlines the fundamental material and political parameters that facilitate or constrain policy change and analyses the demand for and supply of mitigation policy. The fourth section presents possible mitigation policy trajectories. The analysis is based on available documents and previous research, and data from fieldwork in Brazil between August 2013 and February 2014, including 20 semi-structured interviews with key informants and participant observation of climate policy debates.[1]

Table 2.1 GHG emissions and energy data for Brazil[a]

Total GHG emissions by year	1990	2005	2013
	1807 $MtCO_2e$[b]	2349 $MtCO_2e$	1568 $MtCO_2e$
GHG emissions per capita 2013			7.84 tCO_2e
GHG emissions by sector 2013	Agriculture 26.6%, energy 30.2%, LULUCF 34.6%,[c] industrial processes 5.5%, waste 3.1%		
Energy mix 2013[d]	Fossil fuels 57.7%, renewables 41%, nuclear 1.3%		

Notes:

a. Scientists are uncertain and disagree on how to best measure and report emissions from agriculture and deforestation. In this chapter I use the emissions data presented by the Brazilian NGO Observatório do Clima in their Sistema de Estimativa de Emissão de Gases de Efeito Estufa (SEEG) where other GHGs are translated to CO_2 equivalents based on global warming potential. The Brazilian government has not published official data after 2010, so I use SEEG to get recently updated emissions data from all sectors (SEEG, 2014).
b. Million metric tons CO_2 equivalents.
c. Land use, land use change and forestry.
d. Total domestic energy supply, including the energy that is lost between supply and consumption. Data from National Energy Balance (EPE, 2014).

CURRENT CLIMATE POLICY AND RECENT TRENDS

The Climate Law

The reduction in deforestation started in 2005 during President Luiz Ignácio 'Lula' da Silva's first government. A national plan on climate change was adopted in 2008 and a climate bill was passed by both chambers of Congress in December 2009, and was signed by President Lula after COP 15 in Copenhagen, making Brazil one of the first countries with a comprehensive climate law (GoB, 2009). Brazil's voluntary commitment in Copenhagen to reduce GHG emissions by 36.1–38.9 per cent below BAU trajectories by 2020 was included in the law.

The policy processes leading to the Climate Law in 2009 have been thoroughly analysed in previous research. Hochstetler and Keck's 2007 book covers the history, institution building and political conflicts of environmentalism in Brazil, especially emphasizing the role of domestic social movements. Carvalho (2010) follows with a thorough analysis of forest policy processes, linking the domestic and the international levels. Viola has published extensively on Brazilian and international climate politics: Hochstetler and Viola (2012) and Viola and Franchini (2012, 2014) analyse the decision-making process of the Climate Law.

All the above-mentioned find the bargaining between different branches of government important for the climate policy outcomes in Brazil. Hochstetler and Keck (2007) find that environmental activists' entrance in the Ministry of Environment (MMA)[2] during Lula's presidency increased the ministry's possibility to set the agenda and take advantage of opportunities for policy change. The Ministry of Agriculture and the Ministry of Mines and Energy (MME) were reluctant about domestic climate policies and mitigation targets from the start, arguing that targets could harm Brazilian food security and energy security.

Carvalho (2010) emphasizes the increasing importance of the MMA in the government's climate policy formation during Lula's governments when Marina Silva (2003–08) and Carlos Minc (2008–10) were ministers of environment. In the beginning of the 2000s the MMA's role in Brazilian climate politics was limited. Climate policy was largely limited to participation in the international climate negotiations and the Ministry of Foreign Affairs (MoFA) had the main responsibility for policy development, aided by the Ministry of Science, Technology and Innovation (MCTI) for technical support (ibid.). Supported by an environmental coalition of NGOs and climate scientists, Silva played an important role in integrating climate change in the MMA's portfolio (ibid.). The inclusion of forest in the climate policy debate was important for the agenda setting of domestic climate policies.

Hochstetler and Viola (2012) emphasize the importance of a 'bootleggers and Baptists' coalition between industry stakeholders and environmentalists for the climate policy outcome. The coalition was created when two influencing factors coincided during the discussion of the Climate Law in 2009. First, Brazil had already cut deforestation emissions significantly and environmentalists argued that these reductions could be translated into a target. Second, Brazilian business stakeholders thought border tax adjustments in the United States and the European Union would soon be implemented, and therefore saw it as being to their advantage if Brazil had its own mitigation targets and participated in a global climate agreement (ibid.).

Kasa (2013) nuances the focus on domestic actors by emphasizing how initiatives in the international climate change negotiations changed the preferences of Brazilian stakeholders. The implementation of policies to reduce deforestation was facilitated by the inclusion of forest in the international negotiations and the introduction of the REDD[3] and REDD+ mechanisms (ibid.).

A political driver behind the Climate Law that is given less attention in previous studies is the competition between Dilma Rousseff from the Workers' Party (PT) and José Serra from the Brazilian Social Democratic Party (PSDB) before the presidential elections in 2010. Hochstetler and Viola (2012, pp. 763–4) and Viola and Franchini (2012, pp. 189–90) emphasize the importance of Marina Silva's presidential candidacy for the emergence of climate change as an important issue in the electoral debate.

In 2009, however, despite Silva's candidature, Lula and the PT saw Serra as their main rival and disliked his popularity in the polls. Rousseff was then Minister of Cabinet Coordination, but because she had been Minister of Energy and Mines she had low environmental legitimacy. When Serra, as Governor of São Paulo, launched a state climate plan in the autumn of 2009, Lula found it necessary to strengthen Rousseff's climate image. Serra presented his plan on a Monday and the government presented climate targets the following Friday. With Marina Silva in the Senate, the targets were introduced when the climate bill was being processed in the Senate. The government was initially opposed to legally binding targets, but later agreed to include targets in the law.

The decisions to include mitigation targets in the Climate Law and to announce them as voluntary commitments at COP 15 in Copenhagen were made by Lula without full governmental support. The ministries that opposed domestic mitigation targets felt ignored and demanded changes in the law. President Lula agreed to veto the most disputed articles in the Climate Law (Interview 1 – see the Appendix to this chapter). The article protecting climate measures from budgetary cuts was vetoed by Lula after protests from the ministries of finance and planning and the Attorney General. The articles calling for investments in new renewables and a gradual phase-out of fossil fuel energy were vetoed after protests from the MME. The MME argued that these articles were in conflict with energy security goals, in addition to being unnecessary since Brazil already had one of the world's cleanest energy matrices (GoB, 2009). The latter vetoes illustrate the resistance towards mitigation in the Brazilian energy sector.

The Brazilian mitigation targets are based on estimated BAU emissions of 3236 million metric tons CO_2 equivalents ($MtCO_2e$) in 2020 (GoB, 2010). The focus in the Climate Law and the follow-up decree is mainly on deforestation emissions. If these emissions are kept at approximately the 2012 levels, Brazil can reach its emissions target even if emissions from energy and agriculture increase significantly (see Table 2.1). The energy and agriculture sectors have good mitigation possibilities, but the sectorial plans for Climate Law implementation vary in their focus on mitigation possibilities.

Sectorial Plans

In December 2010 President Lula issued a decree on the Climate Law implementation (GoB, 2010). The decree announced five sectorial plans: the plan for deforestation control in the Amazon; the plan for deforestation control in the Cerrado; the energy plan; the low-carbon agriculture plan (the ABC plan); and the steel industry plan. Later, four more mitigation plans were added, for transport and urban mobility, mining, industry and health, but the overall mitigation target of 36.1–38.9 per cent below BAU trajectories was not upgraded. The most important plans from a mitigation perspective are the ones targeting the emissions from deforestation, agriculture and energy.

Deforestation

The Amazon and Cerrado plans are based on the 1965 Forest Code that protects between 50 and 80 per cent of the Amazon forest. These plans were already in the process of being implemented when the Climate Law was adopted. The Forest Code had long been neglected prior to Marina Silva's becoming Minister of Environment in 2003. Through the MMA's intensive work and cooperation with state governments, scientists and NGOs, deforestation control and enhancement of indigenous peoples' control over territories significantly reduced deforestation in the Amazon from 2005 onwards (Kasa, 2013).

In 2012, during Dilma Rousseff's first government, the Forest Code was revised after pressure from agribusiness interests. The revised code fills some regulatory gaps, like requiring rural property registration, but also gives an amnesty for illegal deforestation in the past and reduces the protection of the Cerrado and of forest around rivers and lakes (Viola and Franchini, 2014). The deforestation rate increased by 28 per cent in 2013 compared to 2012 and environmentalists see the increase as an alarming sign (IPAM et al., 2014). It is too early to say whether the increased deforestation is a result of the Forest Code revisions. The development needs to be followed closely during the coming years before one can conclude that stable and low deforestation rates have been achieved (Interview 2).

Agriculture

The ABC plan for agriculture includes both mitigation and adaptation measures. The Brazilian Corporation of Agricultural Research, Embrapa, has been engaged in finding low-carbon solutions. Many of the low-carbon solutions are economically and qualitatively efficient and competitive. All measures under the ABC plan are voluntary, new loans and other incentives are introduced, but there is no punishment for BAU practice. The plan does not include targets (GoB, 2012a; Stabile et al., 2012).

The main hurdles to implementation of the ABC plan are technicalities in the financial incentives; investments for low-carbon agriculture still amount to only around 3 per cent of total agriculture investments (Interview 3; Stabile et al., 2012). Economic interests and food security concerns still dominate Brazilian agricultural policy, but the introduction of low-carbon solutions has started and the trend is positive, although slow. The trajectories of Brazilian agricultural emissions also depend on international demand for agricultural products, especially soybeans and beef (Rodrigues-Filho et al., 2014).

Energy

Energy emissions in Brazil have two main sources: electricity generation and transportation. Two of the sectorial climate plans cover these sources. The energy plan is a mildly modified version of already existing energy sector plans. The transport and urban mobility plan is a merger of two existing plans: the cargo plan for moving more long-distance transport from road to rail and to waterways, and the human transport plan that was launched in connection with the 2014 Football World Cup and 2016 Summer Olympics, not originally a mitigation plan. The interest to innovate and invest in new low-carbon solutions has been miniscule in both energy emission sectors.

The energy plan is based on BAU energy policy trajectories, where fossil fuel use is maximized and GDP growth is estimated at 4.5 per cent annually, much higher than the actual growth. Lower-than-BAU emissions in the energy sector are thus easy to attain (Interview 4). The transport and urban mobility plan is far from being implemented; the plan lacks emission targets, but the plan's measures would lower current and future transport emissions significantly, if implemented (Interview 3).

As a macroeconomic tool to control inflation, the government has introduced a cap on petrol prices. Most cars are flex-fuel and can use any mix of petrol and ethanol, but people prefer to use petrol if the price difference is small. Brazil almost reached self-sufficiency in petroleum in

2008, but imports have since increased because of expansion of the vehicle fleet. The price cap makes reducing transport emissions more difficult.

International Climate Cooperation

As a promoter of sustainable development since the UN Summit in Rio in 1992, the Brazilian government feels like a steward for the UN-based climate regime (Interview 5; GoB, 2012b). Except for the domestic discussion on deforestation, the Brazilian government has mainly treated climate change as a foreign policy issue. Internationally Brazil has pushed for an ambitious agreement, but clearly emphasized that the Annex 1 countries have to take the lead and cut their emissions significantly first since much of the accumulated CO_2 in the atmosphere is from historical emissions (GoB, 2012b).

The past five years Brazil has allied with China, India and South Africa in the BASIC group, defending a continued divide between Annex 1 and non-Annex 1 countries based on the UNFCCC platform of 'common but differentiated responsibilities and respective capabilities'. From a Brazilian perspective, the BASIC countries took great responsibility in Copenhagen in 2009, presenting national emissions targets and voluntary commitments.

Brazil has cut its aggregate GHG emissions more than any other developing country and more than most Annex 1 countries. Still, the Annex 1 countries do not recognize Brazil as a leading contributor and have rejected Brazil's proposals for a global distribution of mitigation responsibility based on historical responsibility.

Brazil's abundant natural resources and its capable scientific community make many mitigation measures less expensive and troublesome in Brazil than in many other countries. An ambitious international mitigation effort could pressure Brazil into making the necessary investments for growth in low-carbon solutions. With high mitigation records and one of the world's cleanest energy matrices, Brazil feels no such international pressure currently.

Nevertheless, Brazil has solid diplomatic traditions and tends to be willing to strike pragmatic deals in international negotiations. From a Brazilian perspective, the ball is now in the court of the Annex 1 countries, but Brazil is likely to become more cooperative again if Annex 1 countries show more willingness to contribute to mitigation.

EXPLAINING POLICY TRAJECTORIES

Fundamental Parameters

Country of diversities and natural abundance

Brazil has had one of the world's most unequal income distributions for decades. Inequality is slowly declining because of poverty reduction, but it is still high. Socioeconomic inequalities have created clientelistic societal structures in poor and rural areas where the wealth of political and economic elites stands in stark contrast to the average wealth of the rest of the population. These clientelistic structures result in disproportionate representation of conservative, office-seeking, and status quo–oriented interests in state and federal decision-making, decreasing the possibility for policy change.

Large geographical and socioeconomic differences exist between Brazil's five regions: the large, poor and scarcely populated north where the Amazon rainforest is located; the economic centre in the south-east with the megacities São Paulo and Rio de Janeiro; the poor north-east with high climate vulnerability; the increasingly affluent centre-west with the Cerrado savannah where deforestation is increasing following a recent soybean production boom; and finally, the south with higher than average living standards. The regional differences influence which interests decision-makers from each region tend to represent and defend in Congress.

Brazil has widely varied and abundant natural resources. The renewable energy potential is great, with large hydropower resources and optimal conditions for wind and solar power in many parts of Brazil. More than 70 per cent of Brazilian electricity is generated from hydropower (EPE, 2014). But because of relatively flat river systems, most of the hydropower is generated through construction of large dams. Dam construction is controversial, especially in forested areas where building the infrastructure needed to construct a dam causes legal and illegal deforestation, and vegetation covered by water produces methane emissions. As much as two-thirds of deforestation in the Amazon is in connection with infrastructure projects such as roads and dams (Interview 6; Vera-Diaz et al., 2009).

Dependency on imported oil has been the Achilles' heel of Brazilian energy security. Following the global oil crisis in 1973, the development of biofuels to substitute imported oil was intensified and ethanol from sugar cane became an important fuel for the domestic vehicle fleet. In 2007 Brazil announced large offshore pre-salt petroleum reserves. The pre-salt oil production has started and is currently increasing. It is expected to grow and make Brazil a net exporter of oil and possibly natural gas after 2020 (Tolmasquim, 2012).

The southern region is the only coal-producing region in Brazil, and although the coal is of low quality, it is argued to be important for energy security in the region, and the southern coal industry is well represented in Congress. New coal plants have also been built for cheap coal imported from Asia. To prevent the expansion of coal in the energy matrix, green politicians in Congress are working to introduce tax regulations to make imported coal unprofitable (Interview 1).

Brazil is the world's largest producer and exporter of sugar, soybeans, coffee, beef, orange juice and tobacco. Land is very unequally distributed, so a large amount of the agricultural resources is owned by a small elite. The interests of this agribusiness elite are among the best represented in Brazilian politics.

Increased Asian demand for commodities has increased the prices and the exports from Brazil, while Brazilian industry strives to compete with cheap Asian manufactured products. Increasing soybean and beef production causes agricultural emissions from soil use and enteric fermentation, deforestation emissions from the expansion of farmland and energy emissions from transportation of agricultural products.

Infrastructure is the most important material hindrance to mitigation. When the governments in the 1950s and 1960s planned modern Brazil, they made strategic decisions to develop road-based transportation infrastructure and hydroelectricity-based energy infrastructure. Incentive structures are still built around these solutions and the companies specialized in road and dam construction, among the largest companies in Brazil, have the political and lobbying power to perpetuate such incentive structures. Since 2000 domestic transportation of goods and passengers has increased rapidly, but corresponding needs for new and improved infrastructure have not been prioritized. The few new investments have concentrated on further development of existing infrastructure, or on new projects involving similar infrastructure. Railways and other low-carbon solutions are scarce.

In addition to transportation, Brazil has other energy-infrastructure challenges. Pipelines for oil and ethanol, and electricity-grid development to include new wind and solar energy are infrastructure investments that would reduce emissions significantly without reducing the quality of energy services, but these investments are not prioritized (Interview 7).

Car ownership is an important cultural symbol of wealth and freedom in Brazil, and the government has prioritized incentives for private vehicles over sustainable public transportation. Safety concerns in many

Brazilian cities and on interstate roads cause people to choose private transport or air travel if they can. This problem is part of a larger structural challenge of violence, crime and corruption that affects all levels of Brazilian politics.[4]

Political system

After a military dictatorship from 1964 to 1985, the current Brazilian democratic constitution was adopted in 1988. The National Congress of Brazil is bicameral, consisting of the Chamber of Deputies and the Senate. Three senators are elected from each state and the Federal District for eight-year terms, and the chamber deputies are elected for four-year terms through a system of proportional representation where small states are relatively better represented than large states.

'Presidentialism by coalition' is used to describe the Brazilian political system. The Brazilian president has extensive power through proposing, signing and vetoing laws; through governing by decrees and provisional measures; and through blocking 'the congressional agenda by requests for urgent appreciation of his proposals' (Santos and Pegurier, 2011, p. 169). However, political parties seldom achieve more than 20 per cent of the seats in Congress, so the elected president has to form a coalition government to get majority support. The necessity to include an array of political parties in the government makes it difficult to get support for major policy changes.

All members of Congress can propose bills and amendments, but the majority of enacted policy proposals come from the government, so the executive is the main agenda setter (Figueiredo and Limongi, 2000, p. 155). Party cohesion in Congress is quite high, so the government can expect majority support after bargaining an agreement with the party leaders in the government coalition (Cheibub et al., 2009). It is not unusual for the president to veto laws or parts of laws that were initially proposed by the president; this was the case with the Climate Law in 2009. Decision-making on state and municipality levels is similar to federal decision-making.

The president, governors and mayors are directly elected for four-year terms and incumbents can be re-elected once. Brazilians that hold executive positions and want to run for another office have to resign from their current position. Federal and state elections are held together, while municipal elections are held in between, so there are elections every second year. These election rules lead to frequent changes in high-ranking offices. New political leadership normally leads to high turnover in administrative positions, such as the approximately 25 000 political appointments that a president can make in the government apparatus (Fleischer, 2011).

This high turnover in administrations at all governance levels causes a tradition of short-term thinking in policy-making and planning. Both politicians and contractors see long-term projects as risky, and there is no stable administrative management ensuring the continuity of projects that span administrations. Mitigation measures that are relatively cheap in Brazil compared to equivalent measures in other large countries can be considered too politically expensive by Brazilian politicians who need to deliver results as early in their tenure as possible.

Lobbying is common and interest groups work hard to get their own representatives nominated and elected for political office (Viola and Franchini, 2012, p. 180). Many politicians are recruited from the elites within important sectors, such as senator from Tocantins, Kátia Abreu, who is also president of the Brazilian Confederation of Agriculture and Livestock (CNA), Brazil's main agricultural interest organization.

Brazilian bureaucracy is status quo–oriented and formalistic with detailed legalization and regulation. But the frequent reshuffle between positions in state and civil society makes politicians open to public demand and creates windows of opportunity for policy change (Hochstetler and Keck, 2007, pp. 225–30). Hochstetler and Keck (ibid., p. 19) emphasize the importance of networks in Brazilian politics; networks of organizations and individuals connect with policy-makers to push their agenda through the political channels and to get inside information from politicians and bureaucrats. Many of these networks of informal and formal policy-makers function as advocacy coalitions.

Brazilian government ministries are important veto players in policy processes. Ministries are usually divided proportionally among government coalition parties and can significantly influence both policy drafting and decision-making. The information presented to government and Congress by a ministry is usually perceived as more legitimate and trustworthy than information from the outside (Interview 4). The bargaining between ministries in the government favours strong ministries that have either strong political clout or a strong technical body representing an important economic sector.

Political parties and alliances

The fragmented political party structure in Brazil leads to considerable bargaining in Congress and government. Following the 2014 elections, 26 parties were represented in Congress. The largest parties are the centre-left Workers' Party (PT), the centre-right Brazilian Social Democratic Party (PSDB) and the conservative and office-seeking Brazilian

Democratic Movement Party (PMDB). PT and PSDB are the main opponents in elections, while PMDB is usually necessary to include in any governing coalition to get majority support in Congress. President Rousseff from PT was re-elected in 2014 with support from PMDB.

There are also cross-party alliances along cleavages that are not reflected in the party structure in Congress. The most important alliance for climate policy development is the *bancada ruralista*, the rural caucus, a cross-party block representing rural interests in Congress. The rural caucus has members from most political parties and defends the interests of the rural middle class and elites such as landowners, farmers and local governments and business people. The rural caucus can usually mobilize a majority in the Senate, and often in the Chamber of Deputies, too. Deal-making and exchange of support between different conservative caucuses in Congress make policy change difficult.

Driving Forces in Brazilian Climate Policy Development and Implementation

Since strong interest groups in society have succeeded in getting their representatives elected, the division between demand and supply is rather blurry in Brazilian policy-making. In mitigation-related policy the blurriness is especially present in energy policy-making where many of the same actors are involved in both demand and supply. 'The energy sectors do not need a lobby; they have the Ministry of Mines and Energy to do all the lobbying for them' (Interview 4).

Societal demand

Climate change is a relatively new issue in Brazilian politics; it was discussed in relation to foreign policy and environmental policy before, but climate change as a policy subsystem, where all actors interested in climate politics operate in the same arena of policy debate, emerged during Lula's first government when Marina Silva was Minister of Environment.

Four advocacy coalitions cover the political interests involved in mitigation policy-making. Their degrees of 'co-ordinated activity over time' (Sabatier, 1998, p. 103) vary, but the coalitions recognize each other as organized blocks of interests in the political debate. The four advocacy coalitions are also active in policy-making in the subsystems of forest and agriculture, energy and transport.

The demand for reduced deforestation has had popular support in Brazil since the 1970s. Biodiversity, indigenous peoples' rights and

Brazil's responsibility to protect the nature the country has been trusted to manage, have been the main arguments against deforestation.

The environmental advocacy coalition emerged when scientists working in Brazilian NGOs started lobbying to get GHGs from deforestation and land use change included in countries' official aggregate GHG emissions. Local, national and international NGOs were the first core members of this coalition (Interview 3). From 2005 to 2010 the environmental coalition was strong and included NGOs, scientists, journalists in leading national news networks, green politicians, business people and people in high administrative and political positions in the government, mainly in the MMA. The coalition was important for connecting core forest policy concerns in civil society to the climate change agenda. The members of this coalition share the core belief that mitigation of environmental degradation and climate change should be prioritized in all relevant policy decisions, and that Brazil has better capacity and opportunity to make such priorities than many other countries do.

The main opponents of deforestation control have been cattle ranchers, the timber industry and farmers in the Amazon and Cerrado. These actors form the agribusiness advocacy coalition together with the bureaucracy in the Ministry of Agriculture and the rural caucus in Congress. The members share the belief that stricter forest and agriculture regulations and control restrict private property rights and jeopardize food security and the possibility to make a living in Brazilian agriculture.

The climate debate has nevertheless served to inform the agribusiness coalition about the impacts of climate change. The coalition is now positive to more research on low-carbon agriculture and on effective production methods to minimize further deforestation, and welcomes such initiatives as long as food security and economic profit are taken into account. The policy process regarding the ABC plan that was presented in 2012 was one of the most inclusive of the sectorial climate policy processes. Large differences remain between the environmentalists and agribusiness regarding designing a sustainable Brazilian forest policy.

The Brazilian energy sector is highly regulated, the core beliefs and concerns in Brazilian energy policy-making have been institutionalized, and there are few public debates on energy policy. The main energy policy decisions are taken on the federal level by the National Council for Energy Policy (CNPE) led by the MME. The energy advocacy coalition includes the MME; the national energy companies Petrobras and Electrobras; large engineering companies such as Odebrecht; the national Energy Research Company (EPE); the leaderships of the National Electric Energy Agency (ANEEL) and of the National Agency of Petroleum, Natural Gas and Biofuels (ANP); and energy politicians from the largest parties in Congress.

Most of these actors are represented in CNPE and share a core belief closely connected to the overall political aim of economic growth and development. That core belief is that energy security is of vital importance to Brazil and that self-sufficiency in energy supply is the best way to attain security. The coalition also emphasizes that Brazil has one of the most renewable energy matrices in the world. According to law, CNPE is supposed to include independent energy experts and civil society representatives, but these actors are currently not represented on the council (Interview 3).

The fourth advocacy coalition is important for mitigation policy development in the transport sector. The coalition – consisting of Brazilian car producers, the car industry labour unions, the Ministry of Transport and the general public (in its demand for cheap cars) – acts as a barrier to mitigation measures in the transport sector. This coalition argues that the car industry is crucial for the Brazilian economy and for regular Brazilians' possibility to gain freedom through buying and driving cars made in Brazil.

The general public demand for mitigation in the energy sector varies. In the transport sector there is strong demand for better and cheaper public transport; millions of Brazilians still depend on public transport to go to work and to school. Better and safer public transport would also make this option more attractive to the car-owning middle class.

Regarding electricity the main public demand is reliable and cheap electricity. Brazilians are proud of the sustainable energy system based on renewable hydropower, but there is little active demand to keep the energy matrix clean. The protests in connection with large hydroelectric dams in the Amazon, like the large Belo Monte Dam in Pará state, have nonetheless attracted some negative attention to new hydropower projects.

The growing petroleum sector was until recently considered well functioning. The national oil company Petrobras was seen as one of the most competent and internationally competitive companies in Brazil. In the autumn of 2014 a huge corruption scandal was revealed in Petrobras, and this will probably affect the coming developments in the Brazilian petroleum sector. Societal demand for a decrease in petroleum activity is very small, and it is uncertain whether the corruption scandal will influence it.[5]

Governmental supply

The main changes in Brazilian climate policy the past decade are the increased deforestation control starting in 2005, and the Climate Law from 2009. Including the revisions in the Forest Code, the developments from 2011 onwards have not been marked by a governmental willingness to enhance mitigation.

Since 2010 the MMA has lost much of its veto power in climate policy-making. The ministry's budget has been cut, Minister for Environment Izabella Teixeira has little political clout, and she has shown little willingness to cooperate with civil society and NGOs (FBOMS et al., 2013). The Forest Code revisions were supported by Teixeira, representing a great setback for the environmental coalition and showing how the agribusiness coalition still has a more institutionalized power basis than do environmentalists (Viola and Franchini, 2014). Former MMA bureaucrats that now work for other ministries or outside the government are still important members of the environmental advocacy coalition, but they have less influence on climate policy-making than before.

The MME is the main veto player in energy policy development. It is very difficult for other ministries or actors to compete with the legitimacy that information from the MME and EPE has in Congress. The MME and EPE were the architects behind the sectorial plan for energy and they were mainly concerned with energy security and less interested in the climate implications of their sector's activities (Interview 2).

The MME has a strong technical and bureaucratic body. 'The people working there are competent and open to discussions, but they are absolutely liberal, they want free competition and low prices, they all think the same and they often win debates' (Interview 4). During Lula's second term and both of Rousseff's terms, the MME has been headed by PMDB, which has a nationalistic and growth-focused view on natural resource development. PMDB is not interested in including climate change concerns in the MME's portfolio (Interview 8).

The government nonetheless aims to keep the Brazilian energy matrix 40–45 per cent renewable (Tolmasquim, 2012). Electricity production concessions are granted after open bidding. In the bidding rounds electricity prices and energy security determine which projects are granted concessions, and the government is unwilling to introduce subsidies to promote new renewables (Interview 9).

With rapid development of new wind power technology in the last decade, wind power plants have won several electricity bids and wind power can often compete with coal and nuclear power at price parity. In October 2014 ANEEL held the first bid for electricity from solar power in Brazil, and new concessions for solar (889.7 MW) and wind (769.1 MW) were granted (GoB, 2014).

Most of the planned new renewable energy is nevertheless new hydropower projects. EPE estimates that Brazil's energy needs will double in the next ten years and the government sees new renewables like wind and solar as insufficient, too instable and too expensive to meet future needs (Interview 9).

High international sugar prices and the petrol price cap have led to a decrease in ethanol production in Brazil, but the government says ethanol and other kinds of biofuel are still a priority. Brazil wants to increase production and export of biofuels and rejects the view that biofuels compete with food production (Interview 10). As long as the possible impact on deforestation is monitored, environmentalists agree that increased biofuel and biomass production is an important mitigation measure for Brazil (Interview 6).

Petrobras, the sugar cane industry, the environmental movement and the energy coalition demand changes in the petrol price cap. If liberal competition principles were implemented, ethanol and other biofuels would be more competitive in the Brazilian market and the share of fossil fuels in transportation and power generation would decrease. But changes in the price policy are blocked by the Ministry of Finance's concern for inflation. Public demand is very sensitive to inflation and price increases, and the government uses the price cap as a macroeconomic tool to control inflation. However, some adjustments in the price cap have been announced to ease the losses for energy producers.

Some Brazilian cities, like Curitiba, have well-functioning public transport systems, but these cities are exceptions. Urban transport is the responsibility of local governments. High turnover in political and administrative positions makes it difficult to implement and follow up large infrastructure projects, and vertical and horizontal corruption in the construction process causes delays and low-quality results.

Supply–demand relationship

The environmental coalition argues that the large potential for wind, solar and biomass electricity in Brazil should be incentivized and given priority in the national bidding rounds for new electricity projects (Interview 11). The predominant view in the government is that Brazil is already mitigating emissions through the large percentage of hydropower in electricity production (Interview 9).

Environmentalists argue that the energy coalition is stuck in its focus on new, large energy projects and is unwilling to take new and innovative solutions into account (Interview 11). Economic interests lead construction companies and electricity companies to favour new, large projects involving hydropower, nuclear power or fossil fuel in their advice to the government. Small wind and solar projects or energy efficiency is less profitable for them (Interview 8). After pressure from green politicians and innovation interests in the MCTI, the first bidding rounds for new renewables took place in 2014.

Environmentalists have mobilized strongly against the Belo Monte Dam, leading to environmental evaluations and large reductions in the original project. Despite the delays, Belo Monte is still being built, but now with significantly reduced electricity generation potential. The government argues that the environmental coalition's resistance to new hydroelectric projects leads to a growth in fossil fuel electricity generation in Brazil (Interview 10). Better implementation of the regulations for environmental sustainability assessments, before new projects are approved, could probably ease the conflict between environmental concerns and new hydro (Interview 11).

Future success and growth in the petroleum sector are connected to political goals of energy independence, eradication of poverty and energy for all, so it has been impossible for the environmental coalition to make climate change mitigation a priority above other concerns in the petroleum sector.

The ties between the government and Petrobras are tight, but the environmental coalition sees the government as a greater barrier to policy change than Petrobras and the other oil companies are. Both Petrobras and the regulator ANP have become less politically independent than was intended when the current sector structure was introduced at the end of the 1990s (Interview 7).

Environmental regulations of petroleum activities are already relatively strict in Brazil, and the Brazilian Institute of Environment and Renewable Natural Resources (IBAMA) works actively with the sector to ensure compliance. The workload on IBAMA delays projects and such delays are a much larger concern for petroleum companies than the environmental and climate regulations themselves (Interview 12). The petroleum companies have finances and capacity to adjust to environmental regulations, and if any future policy leads to less demand for oil in Brazil they can easily sell their pre-salt light crude on the international market. The main threats to an increase in offshore oil and gas production in Brazil are low international oil prices[6] and technical difficulties related to the geological profile of the reserves.

For most Brazilians and the international climate community, mitigation in Brazil is synonymous with deforestation control. Little public attention is given to other sectors. The NGOs that have built their strength in the fight against deforestation are concerned about other emission sectors and want to build capacity to become knowledge providers also in these sectors. It is, however, difficult for them to get national or international funding for such capacity building, and the environmental coalition's possibilities to compete with the energy coalition for public and political support is therefore low (Interview 8).

LIKELY POLICY TRAJECTORIES

Although energy emissions have an increasing role in the Brazilian climate debate, the main political debate in Brazilian mitigation policy during the coming five years is likely to concentrate on deforestation. Deforestation has increased since 2012 and the struggle between the environmental and the agribusiness coalitions will continue.

The veto power of the rural caucus has increased, especially after Kátia Abreu was announced as new Minister of Agriculture from January 2015. But several ministries, mainly the MMA, MCTI and the President's Secretariat for Strategic Affairs (SAE), are working to keep deforestation rates low. Since deforestation rates are closely connected to Brazil's international image, the president will most likely support continued focus on deforestation control. The continued deforestation pressure on the rainforest decreases the domestic climate debate on emissions from other sectors.

Enhanced implementation of low-carbon agriculture measures is likely in the coming years. There is no fierce opposition to such measures and there are possibilities to combine mitigation and adaptation in many agricultural areas, but the transition is slow.

The environmental coalition has enhanced its expertise on energy issues and will be active in emphasizing low-carbon energy investments that have co-benefits for energy security and energy prices. Brazil's main mitigation of energy emissions was based on energy security concerns and took place several decades ago, with large investments in hydropower and biofuels. Increasing energy demand and decreasing precipitation rates will keep energy security on the agenda in the foreseeable future. Some of the new energy needs will be filled by wind and solar power; cheap loans are available for investors, and more bids directed at solar and wind power are likely. But this development is highly dependent on energy prices.

Most of the increased electricity demand will nevertheless be met by domestic and imported hydropower and natural gas. Macroeconomic concerns for economic growth and the strong connections between construction companies and politicians will likely lead to new hydroelectric projects being prioritized instead of modernization and efficiency improvements in existing plants. The government sees natural gas as the best back-up when hydropower production is below average.

With the planned inclusion of three Amazonian cities in the national electricity grid from 2015, the use of diesel-generated electricity in Brazil will decrease significantly, but the grid capacity will be further challenged. Grid development is a hurdle for energy efficiency in Brazil and active government investment decisions and smart-grid developments are needed if Brazil wants to enhance energy efficiency.

The most likely possibility for further low-carbon energy development is within biofuels. Despite decreasing domestic and international attention to Lula's 'ethanol diplomacy' that was launched in 2006, ethanol and other biofuels can meet both the environmental coalition's aim of having Brazil take more international responsibility for climate change mitigation and the energy coalition's aims of enhancements to both energy security and energy self-sufficiency, and low energy prices.

Mitigation in the transport sector will remain the most expensive and complex option for Brazil. Apart from increased use of biofuels and natural gas, few new mitigation measures can be expected before 2020. The government will likely increase the share of biofuels in petrol and diesel, and the pressure on the petrol price cap will grow, so ethanol will probably become more competitive. However, because of the expected increase in car ownership, the emissions will continue to grow in this sector.

Energy policy changes towards prioritizing low-carbon solutions are not likely in Brazil before 2020. Internationally there is no pressure on Brazil to mitigate energy emissions, so mitigation has low priority with the president and the MoFA, and it meets resistance in the MME. Energy policy decisions will be made based on energy security and price concerns, and hydropower, petroleum and biofuels will probably grow, whereas coal and nuclear are unlikely to grow significantly, although new concessions for coal power plants were announced in December 2014. However, Brazil will be open to all new energy sources that are price competitive.

Whether 'a low-carbon economy' means 'an economy that is more low carbon than those of most other countries', or 'an economy that is as low-carbon as possible' is not defined in the Climate Law. For Brazil's economy to meet the latter definition is impossible before 2020, and unless new low-carbon solutions are more highly prioritized within energy and agriculture, meeting the first definition by 2030 might even be impossible.

NOTES

1. Interviewees had the opportunity to consent to the use of names and affiliations, and to screen quotations and references. I would like to thank all my informants for finding time in their busy schedules to meet me and share their experiences, thoughts and information. See the Appendix at the end of this chapter for the list of interviewees referred to in the text.
2. In general, this chapter uses English-language names for Brazilian organizations and institutions, but, in order to avoid confusion, the established abbreviations that correspond to the Portuguese-language names are used.
3. Reduced emissions from deforestation and forest degradation.
4. On Transparency International's Corruption Perception Index Brazil ranks number 72 of 175 countries in 2013, scoring 42 out of 100 points where 0 indicates a very corrupt country and 100 a very clean country.
5. An analysis of the effects of the scandal is outside the scope of this chapter.
6. The effects of the drop in oil prices towards the end of 2014 are still uncertain.

REFERENCES

Carvalho, F.V. (2010), 'A posição brasileira nas negociações internacionais sobre florestas e clima (1997–2010) – Do veto a proposição' [The Brazilian position in international negotiations on forests and climate (1997–2010) – from veto to proposal], PhD thesis, University of Brasilia.

Cheibub, J.A., A. Figueiredo and F. Limongi (2009), 'Political parties and governors and determinants of legislative behavior in Brazil's chamber of deputies 1988–2006', *Latin American Politics and Society*, **51**(1), 1–30.

Empresa de Pesquisa Energética (EPE) (2014), *Relatório Final do Balanço Energético Nacional – BEN 2014* [Final Report of the National Energy Balance – BEN], accessed 8 December 2014 at https://ben.epe.gov.br/.

FBOMS, Observatório do Clima and Rede Mata Atlântica (2013), 'Carta aberta para Comissão Mista Permanente sobre Mudanças Climáticas – CMMC' [Open letter to the Permanent Joint Committee on Climate Change – CMMC], Brasilia, 17 October 2013.

Figueiredo, A.C. and F. Limongi (2000), 'Presidential power, legislative organization, and party behavior in Brazil', *Comparative Politics*, **32**(2), 151–70.

Fleischer, D. (2011), 'Attempts at political reform in Brazil: a never-ending story', Brazil Under Rousseff Task Force, Center for Hemispheric Policy, University of Miami.

Government of Brazil (GoB) (2009), *Law No. 12.187 of 29 December 2009 Establishing the National Policy on Climate Change*, accessed 18 June 2014 at http://www.planalto.gov.br/ccivil_03/_ato2007-2010/2009/lei/l12187.htm.

Government of Brazil (GoB) (2010), *Decree No. 7.390 of 9 December 2010*, accessed 18 June 2014 at http://www.planalto.gov.br/ccivil_03/_Ato2007-2010/2010/Decreto/D7390.htm.

Government of Brazil (GoB) (2012a), *Plano Setorial de Mitigação e de Adaptação às Mudanças Climáticas para a Consolidação de uma Economia de Baixa Emissão de Carbono na Agricultura: Plano ABC* [Sector Plan on Mitigation and Adaptation to Climate Change for the Consolidation of a Low-Carbon Economy in Agriculture: Plan ABC], accessed 18 December 2014 at http://www.mma.gov.br/images/arquivo/80076/Plano_ABC_VERSAO_FINAL_13jan2012.pdf.

Government of Brazil (GoB) (2012b), 'Speech by Minister Izabella Teixeira during the Climate Convention. Doha, Qatar, 5 December 2012', accessed 19 December 2014 at http://unfccc.int/resource/docs/cop18_cmp8_hl_statements/Statement%20by%20Brazil.pdf.

Government of Brazil (GoB) (2014), 'Leilão de energia de reserva contrata 1.658,8 MW de potência' [Reserve energy auction contracts 1658.8 MW], Portal Brasil, 1 November 2014, accessed 4 November 2014 at http://www.brasil.gov.br/infraestrutura/2014/11/leilao-contrata-1-658-8-mw-de-potencia-com-investimentos-de-r-7-1-bi.

Hochstetler, K. and M.E. Keck (2007), *Greening Brazil – Environmental Activism in State and Society*, Durham, NC: Duke University Press.

Hochstetler, K. and E. Viola (2012), 'Brazil and the politics of climate change: beyond the global commons', *Environmental Politics*, **21**(5), 753–71.

IPAM, ISA and IMAZON (2014), *O Aumento no Desmatamento na Amazônia em 2013: um ponto fora da curva ou fora de controle?* [The Increase in Deforestation in the Amazon in 2013: Outside of the Curve or Out of Control?], Joint Report, Brasilia, 7 January 2014, accessed 24 April 2015 at http://www.ipam.org.br/biblioteca/livro/Aumento-no-Desmatamento-na-Amazonia-em-2013-um-ponto-fora-da-curva-ou-fora-de-controle-/736.

Kasa, S. (2013), 'The second-image reversed and climate policy: how international influences helped changing Brazil's positions on climate change', *Sustainability*, **5**(3), 1049–66.

La Rovere, E.L., C.B. Dubeux, A.O. Pereira Jr. and W. Wills (2013) 'Brazil beyond 2020: from deforestation to the energy challenge', *Climate Policy*, **13**(S1), 70–86.

Ray, J. (2009), 'In major economies, many see threat from climate change', *Gallup.com*, 8 July 2009, accessed 23 April 2015 at http://www.gallup.com/poll/121526/Major-Economies-Threat-Climate-Change.aspx.

Rodrigues-Filho, S., R. Verburg, M. Bursztyn, D. Lindoso, N. Debortoli and A. Vilhena (2014), 'Election-driven weakening of deforestation control in the Brazilian Amazon', *Land Use Policy*, **43**, 111–18.

Sabatier, P.A. (1998), 'The advocacy coalition framework: revisions and relevance for Europe', *Journal of European Public Policy*, **5**(1), 98–130.

Santos, F.M.G. and F.J.H. Pegurier (2011), 'Political careers in Brazil: long-term trends and cross-sectional variation', *Regional & Federal Studies*, **21**(2), 165–83.

Sistema de Estimativa de Emissões de Gases de Efeito Estufa (SEEG) (2014), SEEG website, accessed 19 December 2014 at http://www.seeg.eco.br/.

Stabile, M.C.C., A. Azevedo and D. Nepstad (2012), *Brazil's 'Low-Carbon Agriculture' Program: Barriers to Implementation*, IPAM Report, Brasilia: Amazon Environmental Research Institute (IPAM), accessed 24 April 2015 at

http://www.gcftaskforce.org/documents/brazil%27s_low-carbon_agriculture_program.pdf.
Tolmasquim, M.T. (2012), 'The energy sector in Brazil: policy and perspectives', *Estudos Avançados*, **26**(74), 249–60.
Vera-Diaz, M.C., R.K. Kaufmann and D.C. Nepstad (2009), 'The environmental impacts of soybean expansion and infrastructure development in Brazil's Amazon Basin', Global Development and Environment Institute Working Paper No. 09-05, Medford, MA: Tufts University.
Viola, E. and M. Franchini (2012), 'Climate politics in Brazil: public awareness, social transformations and emissions reduction', in I. Bailey and H. Compston (eds), *Feeling the Heat – The Politics of Climate Policy in Rapidly Industrializing Countries*, Basingstoke, UK: Palgrave Macmillan, pp. 175–201.
Viola, E. and M. Franchini (2014), 'Brazilian climate politics 2005–2012: ambivalence and paradox', *WIREs Climate Change*, **5**(5), 677–88.
World Bank (2014), 'Indicators', *World Bank.org*, accessed 8 December 2014 at http://data.worldbank.org/indicator.

APPENDIX: LIST OF INTERVIEWS

1. Alfredo Sirkis, Congressman for the Brazilian Socialist Party (PSB) and Chairman of the Joint Climate Change Commission of the Brazilian Congress, Brasilia, 27 November 2013.
2. Former Advisor, MMA, Brasilia, 26 November 2013.
3. Carlos Rittl, Executive Secretary, Observatório do Clima, Brasilia, 3 December 2013.
4. Carlos Tomé, Legal Advisor, the Brazilian Senate, Brasilia, 4 December 2013.
5. Everton Lucero, Director of Division of Climate, Ozone and Chemical Security, MoFA, Brasilia, 2 December 2013.
6. Paulo Moutinho, Executive Director, Amazon Environmental Research Institute (IPAM), Brasilia, 20 November 2013.
7. Jacqueline Mariano, Regulation Specialist, ANP, Rio de Janeiro, 9 December 2013.
8. Adriana Ramos, Vice Executive Secretary, Instituto Socioambiental (ISA), Brasilia, 21 November 2013.
9. Brazilian Senior Official, Brasilia, 6 December 2013.
10. Brazilian Diplomat, MoFA, Brasilia, 5 December 2013.
11. Tasso Azevedo, Social Environmental Entrepreneur and former Advisor, MMA, Rio de Janeiro, 28 January 2014 (telephone interview).
12. Advisors in an International Petroleum Company, Rio de Janeiro, 10 December 2013.

3. China: every day is a winding road

Iselin Stensdal

BOX 3.1 KEY FACTS ABOUT CHINA

Population 2013	1.36 billion
Population growth rate 2013	0.5%
Population density 2013	145 per km^2
GDP per capita 2013	6807.43 USD
GDP growth rate 2013	7.7%
GINI index 2011	37.0
Access to electricity	99.7% of the population
Land area	9 388 211 km^2
Political system	Unitary state
Subnational entities	22 provinces, five autonomous regions (Guangxi, Inner Mongolia, Ningxia, Tibet, Xinjiang), four municipalities directly under the centre (Beijing, Chongqing, Shanghai and Tianjin), two Special Administrative Regions (Hong Kong, Macau)

Source: World Bank (2014).

INTRODUCTION

The People's Republic of China is the world's largest CO_2 producer and has the largest population and second-largest economy (see Box 3.1 and Table 3.1). China is at a transformative stage. The speed of development since the 1980s has put China in a very different situation than 30 years ago. Energy- and emission-intensive industries were the previous locomotives of economic growth. China is aiming to rely on the service sector and more advanced industries in the future. However, comprehensive change takes time.

China is a unitary state, with the Chinese Communist Party (CCP) in control of important government positions, effectively ruling the republic. The greenhouse gas (GHG) mitigation policies are mainly supply driven: the central government provides local governments and communities with measures to execute. The central government's reasons for instructing mitigation policies are many. China's vulnerability to impacts of climate change is an incentive to reduce its anthropogenic emissions. Limited domestic natural resources make efficient use of available resources crucial. Decoupling economic growth from energy consumption and GHG emissions is also beneficial for reducing other forms of pollution.

China's CO_2 emissions increased sharply from 2002 onwards (IEA, 2014b, p. 48). Per capita emissions were for a long time under the world average. China surpassed the average in 2006 and is now somewhat above. China's mitigation policies are gradually becoming more ambitious. Among the recent measures are caps on both coal and energy consumption in 2020, as well as a national emissions trading system (ETS) from 2016–17. The coal and energy caps allow for consumption growth, but on a more restricted scale. Contrary to the dynamic change domestically, at the international UN Framework Convention for Climate Change (UNFCCC) negotiations China is more conservative. Despite arguing to maintain as much as possible of the Convention's developed–developing countries divide in coming years, China has on its own initiative taken steps to aid less-developed developing countries.

Because of growing attention to severe air pollution, China is currently moving faster away from coal use than before. For the central government it is not a question of not wanting less pollution and emissions, but more a question of feasibility in making economic development and environmental protection compatible. It is unlikely that China's emissions will peak before 2020; the government estimates that they will peak around 2030. Until then, mitigation policies contribute to restraining emissions increases.

DEPENDENT VARIABLE: CLIMATE CHANGE MITIGATION POLICIES

China's energy consumption increased rapidly from 2002, as did its GHG emissions. Since 1980, energy efficiency had continuously improved year by year. From 2002 to 2005 the trend reversed because of a surge in coal consumption and energy-intensive industries such as steel (Andrews-Speed, 2012, p. 16). This situation, combined with awareness of China's

Table 3.1 GHG emissions and energy mix in China

Total GHG emissions by year	1990	2005	2012
	2269 $MtCO_2e$[a]	5116 $MtCO_2e$	8206 $MtCO_2e$
GHG emissions per capita 2012			6.1 tCO_2e
GHG emissions by sector 2012	Electricity and heat production 50%, manufacturing and construction 31%, other energy industries' own use 4%,[b] other sectors 7%, transport 9%		
Energy mix 2013	Coal 66%, oil 18.4%, renewables and nuclear 9.8%, natural gas 5.8%		

Notes:

a. Million metric tons CO_2 equivalents.

b. Includes emissions from own use in petroleum refining, the manufacture of solid fuels, coal mining, oil and gas extraction and other energy-producing industries.

Sources: International Energy Agency (IEA); National Bureau of Statistics of China.

vulnerability to climate change and severe pollution, made the central leadership decide that a change of China's development direction was necessary. The State Council released China's National Climate Change Programme in 2007, and climate change has been a national priority in politics since.

The socioeconomic Five-Year Plans are the key policy documents in China. The current one (the 12th, 2011–15), followed up on restructuring the economy; it is the greenest to date, devoting a whole chapter to green development. Here, fighting climate change and saving energy were two important measures in improving development. The Five-Year Plans cover many areas besides green development, such as economic growth targets and forecasts of population growth. Two of China's most important climate change and energy policies that influence reducing GHG emissions are those concerning the macroeconomic intensity targets for carbon emissions and energy use. China pledged in 2009 at the UNFCCC COP 15 in Copenhagen, Denmark that it would reduce its carbon intensity – the amount of carbon emitted per produced unit of GDP – by 40–45 per cent in 2020, compared to the 2005 level. This pledge was incorporated into the subsequent 12th Five-Year Plan with a target of a 17 per cent reduction within the plan's timeframe. Continuing efforts from the 11th Five-Year Plan to reduce energy intensity – amount of energy used per produced unit of GDP – the 12th Five-Year Plan targeted a 16 per cent reduction between 2011 and 2015. Such targets in the Five-Year Plans are labelled either 'guiding' or 'binding', according to how important the government considers fulfilling the targets to be. Both intensity targets are binding in the 12th Five-Year Plan. These intensity targets are essentially about decoupling GDP growth from energy use and carbon emissions. As such, reduced carbon intensity does not necessarily equate carbon-emission reductions in total terms. All countries need positive economic growth, and the Chinese government has found intensity targets most suitable at this development stage. For decoupling economic growth from carbon emissions, carbon market mechanisms were a new policy tool added in the 12th Five-Year Plan. The plan contained a goal of starting with ETS pilots, later to be expanded into a national market. By June 2014, all seven systems had started operations. The pilots together constitute the world's second-largest ETS market, after the EU ETS.

In limiting GHG emissions, controlling coal use and total energy consumption, as well as replacing fossil energy sources with non-fossil sources are essential. In the current five-year period, reaching a non-fossil fuel share of 11.4 per cent is a binding target. In the Strategic Action Plan for Energy Development 2014–20, the target is 15 per cent non-fossils by 2020. Here it is also stated that natural gas should account for at least 10 per cent of the primary energy consumption by 2020, up from around 5 per cent currently (NDRC, 2014a, p. 5). In the China–US climate agreement of November 2014, China committed to reach about a 20 per cent share of non-fossils by 2030, and also to have its CO_2 emissions peak around 2030 (*China Daily*, 2014). The energy plan furthermore puts forward that by 2020 coal should account for no more than 62 per cent of primary energy consumption. The plan also states that China's primary energy consumption in 2020 should be around 4.8 billion tons of standard coal equivalents. This means that the annual increase in energy consumption must be less than 3.5 per cent up to 2020 (NDRC, 2014b).

To restructure the economy such that economic growth is decoupled from GHG emissions, it has been decided in the 12th Five-Year Plan that China will focus on seven select strategic emerging industries. Among these new industries are the energy-saving and environmental protection industry, the new-energy industry with a focus on non-fossils, and the new-energy vehicle industry such as electric and hybrid vehicles; all are industries related to mitigation. The National Plan to Address Climate Change 2014–20 states that these industries should make up around 15 per cent of GDP in 2020, and that the service sector should make up more than 52 per cent of GDP the same year. By comparison, the service industry made up around 43 per cent of the GDP in 2010 (NDRC, 2014a, p. 7; National Bureau of Statistics China, 2014, p. 2.1). Several laws support the turn to a lower-carbon economy, such as the Energy Conservation Law (1997, 2007), the Renewable Energy Law (2005) and the Environmental Protection Law, where the 2014 revision strengthened authorities' tools for ensuring compliance (1989, 2014).

Simultaneously to China's becoming the world's largest GHG emitter, international attention towards China has increased. The attention culminated at the international UNFCCC negotiations in 2009, at COP 15 in Copenhagen. Afterwards several parties singled China out as *the* laggard that was hindering progress in the negotiations. The negotiations are, of course, not only dependent on China, so that is at best only partially true. Nevertheless, being scapegoated was so uncomfortable for China that it has tried to avoid being singled out at COPs since. From 2010 onwards, Chinese statements have expressed greater acceptance of developing country participation. With adoption of the 'Durban Platform' in 2011, China supported that under the new agreement succeeding the Kyoto Protocol, developing countries would also contribute to mitigation efforts. In recent years, as the Alliance of Small Island States (AOSIS) and the European Union came together to drive negotiations further, the contradictions within the traditional developing country block G77 and China have become even more apparent. While pointing to its status as a developing country, China acknowledges that there are differences between developing countries. It has taken a more active role outside the UNFCCC negotiations. For example, China has initiated South–South cooperation, helping less-developed developing countries. The 2014 China–US agreement is another example of an initiative outside the UNFCCC. On the other hand, China has since 2009 also been one of the founders of the Like-Minded Developing Countries (LMDC) negotiation group. The common interest for this mix of developing countries seems to be maintaining much of the preferential developing country treatment from the Convention and the Kyoto Protocol in the future agreement to be effective in 2020.

EXPLAINING POLICY TRAJECTORIES

Fundamental Parameters

Material parameters

China is the world's most populous country and has the world's second-largest economy. It consumed around 19 per cent of the world's energy in 2012, roughly the same amount as the rest of Asia and the African countries combined (IEA, 2014b, p. 30). There are large differences between its provinces in terms of population, economic development, energy consumption and energy endowments (Bian et al., 2013, p. 969). The coastal provinces are generally the richest, most developed and the largest energy consumers.

China is rich in many energy resources. Most abundant is coal, which is also the pillar of China's energy consumption, constituting between 67 and 76 per cent of its consumption between 1980 and 2012 (National Bureau of Statistics of China, 2014, p. 9.2). China also has oil resources, and in 2013 was the fourth-largest oil producer, producing around 5 per cent of the world's crude oil. Since 1993, however, China's oil consumption has exceeded its production (Andrews-Speed, 2012, pp. 14–15; IEA, 2014b, p. 11). Economic growth from the 1980s to 2000s caused its energy consumption to rise. Consumption surged in the 2000s and is mirrored in China's GHG emissions trend. Nonetheless, China has become much more energy efficient, and the renewable energy world leader. From 2001 to 2012 wind power production increased from 749 GWh to 95 980 GWh, solar photovoltaic (PV) production rose from 31 GWh to 6354 GWh, and hydropower production increased over 300 per cent (IEA, 2014a). China aims to be self-sufficient for 85 per cent of its energy needs by 2020, according to its Strategic Action Plan for Energy Development 2014–20 (NDRC, 2014b). Energy-saving and efficiency measures are crucial for reaching this goal. China is especially restricted by its large coal endowment in combination with goals of increasing consumption shares of natural gas and non-fossil energy sources.

China's ecosystems are vulnerable to climate change. The impacts vary from place to place. Around the Three Gorges area, for example, where China has built one of its largest dams, intensified extreme weather conditions trigger soil erosion and harm the ecosystem. There is also an increased risk of floods. Yunnan Province, bordering Myanmar and Vietnam in the south, experienced an extreme drought in 2010. Because of the observed impacts, and the realization that the consequences of climate change in essence may harm the economy, the government emphasizes climate adaptation. In the above-mentioned areas, adaptation capacity building has been ongoing since 2007. In Yunnan Province farmers were given help to climate-proof their winter crops of 2008 so that they did not lose their livelihood (Cao et al., 2012, pp. 67–9).

Institutional and political parameters

The People's Republic of China's formal government structure has these branches: the legislative: National People's Congress (NPC), the executive: State Council, and the judiciary: Supreme People's Court and Supreme People's Procuracy. The National People's Congress (NPC) is the highest organ of power according to the Constitution. In reality, however, the Chinese Communist Party (CCP) effectively rules the republic. CCP members hold key government posts, making the General Secretary of the Party the state's president and the epitome of power. The Standing Committee of the CCP's Politburo is in effect higher hierarchically than the State Council. The role of the state and its organizations is to implement CCP policies (Donald and Benewick, 2005, pp. 56–7; Saich, 2011, p. 142).

China is a unitary state, and below the central (top) government level there are 22 provinces, five autonomous regions and four municipalities. In addition, Hong Kong and Macau are so-called Special Administrative Regions operating under a 'one country, two systems' concept. The provinces, autonomous regions and municipalities in turn have three administrative levels: prefectures, counties/cities and townships/districts. The function and status of prefectures vary greatly from place to place, but the organizational structures of the province, county and township levels mimic that of the central government. The Ministry of Environmental Protection (MEP) at the central level supervises Provincial Environmental Protection Bureaus and so on, for example. The CCP structure is also organized with corresponding entities at each governmental level (Saich, 2011, pp. 179–82).

With the move from a planned economy to a more market-based economy, the state has reduced its grip on society and assumed a more supervisory role in many areas. It still retains a firm grip on issues considered to be of great importance to China's development and stability, like energy. The altering of the bureaucratic structure is in part due to the hefty expenses that the large government apparatus incurs and is therefore a necessity, but also shows the Chinese leaderships' willingness to adapt to new circumstances (Saich, 2011, p. 159; Andrews-Speed, 2012, pp. 7–8).

The central government's energy and climate organizations As for energy and climate organizations, both the energy and climate issues are delegated to the same organization, the macroeconomic planning National Development and Reform Commission (NDRC). Energy has a longer bureaucratic history than climate change does. The large state-owned energy enterprises originally were each organized as a ministry, but were converted to corporations during the 1980s. China had a Ministry of Energy for a few years in the 1980s–1990s, but the current placement with the NDRC occurred in 1998 (Chen, 2012, p. 33; Downs, 2007, p. 70). In the late 1980s and early 1990s climate change was regarded both as a technical issue for scientists, which the China Meteorological Administration managed, and an internationally negotiated issue for the Ministry of Foreign Affairs. As the perception changed to one of regarding climate change as a development issue, the NDRC was mandated responsibility for the issue in 1998 (Stensdal, 2014, pp. 118–19).

NDRC has the sole responsibility for domestic climate change policies, but shares the responsibility with the Ministry of Foreign Affairs for the international UNFCCC negotiations. The energy- and climate-relevant entities of NDRC are the Climate Change Department and the National Energy Administration (NEA). These two entities run the day-to-day management of their corresponding areas, and also serve as secretariats for respective leading groups: the National Energy Commission (NEC) and the National Leading Working Group on Addressing Climate Change (NLWGACC). Forming leading groups on important issues with representatives from relevant ministries and departments is a common practice in China. However, these leading groups are not part of the government structure as prescribed by the Constitution. In 2013 NEC had 22 members and NLWGACC 33, all ministers or directors of their respective organizations (Government of China, 2014a, 2014b). There is overlap between the groups in that both groups are headed by Premier Li Keqiang and Vice-Premier Zhang Gaoli. Fifteen of the representatives serve in both groups, such as the ministers for foreign affairs, science and technology and environmental protection, suggesting the possibility for coordination of policy change.

A practice that affects policy change is that policies adopted are generally unanimously outwardly supported. The power to declare new national policies is limited to the very top State Council, but before a decision is made there are rounds of negotiations, both horizontally, for example, among the ministries, and vertically within the government (Saich, 2011, p. 265). Bargaining is not unique to China; it occurs in most political systems. What is distinctive about policy change in China, however, is the number of organizations and people that must consent to move forward (Lampton, 1992, p. 35). The negotiations slow the speed of policy change. Below the very top leadership, the different ministries and organizations have varying interests and each entity also often finds that it must bargain both horizontally among organizations of the same hierarchical rank and vertically up and down the hierarchy. Negotiation is the tool with which to carve out the best possible situation for where one stands (Lieberthal, 1992, pp. 1–2).

Driving Forces

Governmental supply: the top leadership

China's turn towards lower-carbon development is largely top-down governmentally driven. The top leadership decided already in 2004 that energy and resource saving should be incorporated into the economic restructuring. The backdrop was increasing oil import dependency, coupled with energy shortages in 2003 and 2004. Energy security and natural resource scarcity are two reasons for the shift towards more non-fossil energy sources and energy diversification. Hallding et al. (2009, pp. 125–7) illustrate how the climate change issue really came to the attention of the top leadership around 2005–06, and how quickly they responded by placing climate change at the front of the political agenda with the release of the National Climate Change Programme in June 2007.

Energy and climate policies, of course, do not exist in a vacuum; other interests and issues at hand also play a part in the decisions and can override the importance of climate or energy concerns. Social stability is of utmost importance. Continued economic development and creating jobs are means to a harmonious society. Parallel with the development of energy policies and later climate polices, rural policies have, for example, been a prioritized policy area (Saich, 2011, pp. 271–81).

Since 2013 the government has also instituted a series of policy measures such as overall plans and more hands-on measures to curb severe air pollution. These policies are due to concern for health, but most will also benefit GHG mitigation. While some policy measures, such as shutting down factories, only carry a cost, the government seems keen to find win-win policy options. The establishment of strategic new-energy industries and the ETS each cater to more than one concern;

in addition to mitigation benefits, green jobs can bring continued economic development and possibly a market advantage internationally. The ETS will create a new market estimated to be worth between US$10 and 65 billion by 2020 (*Sydney Morning Herald*, 2014).

The top leadership in China might seem outwardly unified, but personal interests and spheres of influence are common (Saich, 2011, p. 110). The policy changes supplied by the top leadership are informed by expert advice and input from ministries and provinces. The carbon intensity target was, for example, proposed by the National Advisory Committee to the NLWGACC, and is likely to have been first proposed by researchers at Tsinghua University (Wübbeke, 2013, p. 729).

The central bureaucracy Hierarchically closest to the State Council, the ministries, commissions and administrations largely advocate on the basis of their administrative areas. As energy conservation and emission reductions have become two of the top national priorities, it is in this context the group manages its designated policy areas and promotes its preferences to the top leadership for support and resources.

NDRC's central role in day-to-day management suggests a stronger prevalence of development and economic interests in policy input demand (ibid., p. 713). Moreover, the NDRC itself is not a unitary actor; there are differences among the branches and departments. Given the crucial importance of energy to China's economy, the NEA has both more funding and, in cases of diverging interests, more influence than the Climate Change Department. It is, for example, unlikely that the Climate Change Department could launch energy-saving policies that the NEA opposed.

The MEP and the Ministry of Science and Technology (MOST) are favourable towards China's taking an initiative on mitigation actions, in line with economic growth targets (Heggelund et al., 2010, p. 241). The Ministry of Finance has conducted a macroeconomic study with NDRC of low-carbon development towards 2020, 2030 and 2050 (NDRC, 2013, p. 7). While the ministers and administrations in general have different degrees of clout when negotiating among themselves or upwards, sometimes one individual can influence policy change. Wan Gang, Minister of Science and Technology, has pushed for making the automobile industry part of the low-carbon transformation, advocating public transportation as a low-carbon pilot area since 2009. As the new-energy vehicles industry was designated as strategic in the 12th Five-Year Plan, his efforts have been successful. Wan Gang's preference for vehicles can be understood in light of his career. Starting in the 1990s he spent more than a decade working in the automobile industry, before moving into fuel research and eventually into the MOST (*CCTV News*, 2013). In 2013 the MOST conducted a pilot project in 25 cities called '10 cities, 1000 green cars' (NDRC, 2013, p. 23). Arriving at the NPC sessions in 2013 in an electric car with the slogan 'zero-pollution, zero-emissions', Minister Wan later at a press conference stated that 'Colleagues of the ministry are more willing to take EVs [electric vehicles], as every kilometer they travel means a reduction of PM2.5 emission,[1] compared with gas vehicles' (*CCTV News*, 2013). Building a new-energy vehicle industry fits nicely with the objective of incorporating air pollution, energy and climate concerns into continued economic growth. Wan's promotion of the industry is likely a partial explanation for the choice, illustrating that individuals can influence policy choices and changes.

The ETS pilots: businesses' room for (in)action One case that illustrates that policy change is largely top-down driven is that of the ETS pilots, started in 2013 and 2014. The industry sector accounts for around 70 per cent of the energy consumed in China. The seven local governments have had trouble enforcing compliance by the approximately 2000 energy-intensive organizations included in their pilots. In Guangdong Province's pilot, 64 of 242 businesses, mostly cement and steel producers, had not bought any permits to cover their emissions in 2013 when the last auction was closed in May 2014. The enrolled companies had to buy 3 per cent of their permits for 2013, and would then be given the remaining 97 per cent permits for free. One official from an iron and steel producer told Reuters that 'There is no reason for companies to pay millions of yuan a year (for carbon permits) when environment and energy regulators already charge us other pollution fees' (Chen and Reklev, 2014c). Fees for sulphur dioxide emissions, wastewater and solid-waste discharge are among the other fees companies must pay. The China Iron and Steel Association (CISA) argued that the industry's large- and medium-sized steel mills lost RMB 2.3 billion in the first quarter of 2014, partly because of pollution fees. Guangdong's provincial government, on the other hand, had few coercive means. It could impose a fine of up to RMB 50 000 on each company, and could reduce the number of permits for the coming year (ibid.). However, with a permit price ranging between RMB 60 and 72 at the May 2014 auction, paying the fine of RMB 50 000 would be a cheaper option for the companies than buying the full number of permits, even given the prospect of receiving fewer permits the following year. The Guangdong government responded by setting a new deadline to July 2014 and setting up an extra round of auctions in June to give the non-compliant firms a chance to make amends (Chen and Reklev, 2014b).

Societal demand

Local governments From 2007 onwards, provinces were asked to form leading groups on climate change, similar to the national leading group. Since then, city governments have also created groups and task forces to deal with climate change. The provinces and cities have since also formulated their own action plans on climate change (Ye et al., 2008). National policies are implemented locally in the provinces, but delegation of targets is preceded by negotiations between central and provincial governments (Saich, 2011, p. 265). All provinces have individually delegated energy intensity targets, but they have varying strength when bargaining with the central government (ibid., pp. 182–3). The policy is further negotiated between the provincial and the lower levels of government. Reducing energy intensity has been a policy delegated to the local governments since 2006. The provinces' and local governments' influence is strongest in the implementation stage of the policy process. Here they can take on a role similar to that of veto players, by not immediately executing designated policies or skewing the implementation to fit local needs. Resistance to implementing costly designated targets and measures is often linked to local government revenues. Local governments are adept in protecting their own interests when negotiating with a higher-level government organization or a governmental agency at the same hierarchical level.

The energy intensity policy is very important to the central government, and local governments therefore put more effort into implementing it than they do implementing other less emphasized policies. When enforcing implementation of policies at local enterprises, local governments sometimes use bundling strategies, connecting the top-down designated energy policy with local political preferences. Closing small inefficient coal mines will, for example, help the energy intensity as well as increase the extraction rate of the jurisdiction. Another strategy is to align the interests of the enterprises with the interests of the energy-saving policies. Local officials explain to business owners about official rules and requirements, potential benefits of complying as well as the punishment measures that can be used against non-compliant entities. Often the threat of punishment and knowing that any company that does not comply will be punished are enough to convince the enterprises that it is in their own interest to comply. Local governments use different formal and informal incentives towards state-owned enterprises (SOEs) and private corporations. Political incentives can be used for SOE managers, while for private businesses, staying on the good side of the local government can be incentive enough to comply (Kostka and Hobbs,

2012, pp. 766–80). The leaders in Shanxi Province, for example, expected that some businesses would fail to deliver reduction, and created some margin by setting a more stringent target for the province than their designated one from the top (ibid., p. 771).

Some local governments also show supportive demand for low-carbon policies. The appointment of low-carbon pilot cities and provinces in 2010 and 2012 and ETS pilots were supported and wanted by the pilot governments themselves. Also, Qingdao city government has on the mayor's initiative decided to start with ETS in 2015, earlier than the formation of the future national market (Chen and Reklev, 2014d).

Businesses and industries Relevant businesses and industries, either as energy producers or as major energy consumers, are not a unitary actor group; rather, they vary in size and ties to the government. Large energy companies, in particular national oil corporations (NOCs), have become sources of revenue to the government through tax payments (Andrews-Speed, 2012, p. 170). Further empowering certain businesses, some SOEs have administrative status higher than that of local governments, which skews the power balance between the SOEs and the local government responsible for enforcing policies. Under the Beijing ETS pilot, not all companies covered complied willingly to submit 2013 emissions data. One company's employee told Reuters that since the instructions had not come from the highest-level authority, but from the Beijing Development and Reform Commission, they would not comply. This refusal shows that the company felt confident enough to ignore instructions from the local-level government agency (Chen and Reklev, 2014a). Similar to the local governments, businesses can express their policy demand by influencing the executive stage of the policy process, as the ETS pilots illustrate. However, for business leaders, having a good relationship with local leaders and other officials is also valuable. Non-compliance happens when there is room for it. Energy-saving policies can also present opportunities. For large companies, realizing high efficiency standards will in the short term lead to cost increases, but can in the longer run increase their market share by squeezing smaller competitors out of the market (Kostka and Hobbs, 2012, p. 773).

Public opinion: concern for pollution, not climate The decades of economic development have improved most people's lives, but also created big differences between the rich developed cities and the poorer rural areas. There is a visible discrepancy between the welfare systems of urban and rural areas, which the central government is trying to even out (Saich, 2011, pp. 297–304). Climate change is not high on the agenda of most citizens in China. Other things matter more, such as inflation, corruption and the gap between rich and poor. However, concern for the environment has grown in the last five years. Climate change might not be something that concerns the average Chinese citizen, but as schools, highways and airports are closed because of air pollution, everyone becomes aware of the hazards of PM2.5 and the smog. In a 2008 survey, 31 per cent of the public were very concerned about air pollution and 28 per cent about water pollution. By 2013 the number of concerned citizens had risen to 47 per cent and 40 per cent, respectively (Pew Research Center, 2013, p. 2). The public awareness about PM2.5 has happened in only two years; from being a specialized term in 2011, it became a commonly known phenomenon in 2013. The first civilian lawsuit against the state over air pollution came in 2014. A man in Hebei Province filed a lawsuit against the Municipal Environmental Protection Bureau to make it perform its duty of air pollution control, and sought economic compensation for health damage and economic loss (Wee, 2014). When the public's demand for policies addressing their concerns is strong and explicit, it can impel policy change, as the next paragraphs illustrate.

Air pollution: smog gets in your eyes The price of the last three decades' economic wonder and poverty reduction has been environmental degradation; the air quality has become extremely poor in several places. In China, air pollution kills an estimated 670 000 annually (Li, 2014). Much of the local air pollution is caused by burning energy sources such as coal, both industrial and household burning. Heavily polluted air is nothing new; Beijing's air was ranked the third most polluted in the world in 1998 (Andrews, 2008). However, not until 2013 did the State Council bring it to the forefront of the political agenda by announcing a ten-measure action plan to improve China's air quality. What happened?

When Beijing was awarded the 2008 Olympics in 2001, attention towards its air pollution increased. Air pollution PM10 data[2] became publicly available already in 2003, without catching the general public's interest (ibid.). While measures were implemented so that the air quality improved during the Olympics, the pollution remained and increased afterwards. Pre-Olympic measures, such as temporarily terminating factory operations around Beijing, were clearly only short-term solutions.

Joining Twitter in 2008, the US Embassy in Beijing started to tweet hourly PM2.5 measurements. The reporting gained a large following amongst Beijingers who managed to access Twitter despite its being blocked in China. There are several Chinese alternatives to Twitter in China, called Weibo, or microblogs. Air pollution is a frequent topic of the Weibo comments and discussions. In October 2011 Beijing's air became heavily polluted. One day the US Embassy measurement was 'beyond index', that is, above 500, the end of the scale. The situation was much discussed on Weibo. Some Weibo users posted the US Embassy measurements on Weibo, sharing the information with netizens outside Twitter. In January 2012 the Beijing municipal government also started reporting PM2.5 levels (Spegele, 2012). By February 2012 the State Council announced revisions of the air quality standards to also include PM2.5, and to expand monitoring to several cities and areas. Shanghai municipality and other city governments followed suit and started publicizing PM2.5 measurements in 2012 and 2013. The State Council move was discussed on Weibo; on one platform – Sina Weibo – more than a million comments were made (Xinhua, 2013). While there were posts of support for the government's efforts, others bluntly criticized them.[3]

Another PM2.5 record was set on 13 January 2013 in Beijing, exceeding 700. There was little visibility because of smog, flights were delayed and people were advised to stay indoors. By then even state media criticized the government, and the discussion continued on Weibo, many expressing dissatisfaction with having to live in such conditions (Martina, 2013). This was the backdrop for President Xi Jinping and Premier Li Keqiang's accession to their positions during NPC's annual meeting in March 2013. In September 2013 the State Council issued a ten-measure action plan to control air pollution (MEP, 2013). Six months later, at the annual NPC meetings in March 2014, Premier Li Keqiang declared 'war on pollution'. After the national plan was announced, several provinces announced their own goals and targets.

The development in air pollution policies exemplifies that societal demand for policies can influence the central governments' supply. A co-benefit of most measures to reduce air pollution is that they also will curb GHG emissions. Air pollution was also severe in the 1990s and 2000s, but it was not until the frank discussions on Weibo of the record high pollution periods that the central government put the issue on the forefront of its agenda.

A climate-change advocacy coalition? The advisory role of researchers and NGOs learning by cooperation The two actor groups proposing the most explicit policy measures that China can adopt to increase its mitigation efforts are researchers and environmental NGOs (ENGOs). In 2003 the concept of 'scientific outlook on development' was introduced under the leadership of President Hu Jintao and Prime Minister Wen Jiabao. Affirming science's guiding position in policy-making for future development, the concept was added to the Constitution in 2008 (Hallding et al., 2009, p. 124). Researchers at a handful of institutions are consulted in the policy formation process for climate change and energy; their policy demand is requested by the government. The China Meteorological Administration, for instance, regularly reports to politicians on scientific findings. NRDC sometimes takes the initiative to invite researchers from organizations such as the China Academy of Social Sciences (CASS), NDRC's own Energy Research Institute (ERI) and Tsinghua University to compose climate policy proposals. The NDRC's Climate Department reviews proposals, which also can come from ministries or provincial governments, and makes a selection that is presented to the NLWGACC. An expert group is chosen, the National Advisory Committee on Climate Change, varying from ten to 30 members, to advise NLWGACC members on policy options (Wübbeke, 2013, pp. 713–28). Policy suggestions are deliberated in the NLWGACC, and agreed-upon suggestions are put forward as policy proposals to the State Council. Then the State Council instructs the ministries, bureaucracy, provincial and local governments to implement the policy.

Several ENGOs, international and Chinese, have included energy use and climate change in their work areas. The World Wildlife Federation has been operating in China since 1980, but many other international ENGOs, such as the Climate Group, the Energy Foundation, Greenpeace and the World Research Institute, opened offices in China during the 2000s. Chinese ENGOs working on energy and climate include Friends of Nature, Global Environmental Institute, Global Village of Beijing and China Youth Action Network. The ENGOs as a group are not very visibly influential in Chinese policy-making, yet their efforts should not be dismissed. If they have the funding, international ENGOs often pay for projects coordinated by a Chinese ENGO in collaboration with local governments and universities. One example is the Global Environmental Institute's project, started in 2007, which included participation from researchers, businesses, government organizations, ENGOs and financing institutions with an aim to find market-based responses to increase the renewable share in Guangdong Province's energy mix and improve energy efficiency (Stensdal, 2014, pp. 114–22). ENGOs have also teamed with foreign governments to inform and exchange knowledge about low-carbon measures. The workshops on carbon trading, hosted by the Climate Group, with participants from the European Commission and the NDRC, are one example (Climate Group, 2011). The ENGOs' strategy of cooperation to motivate change lays the foundation for building new policies. The involved researchers and ENGOs have had some influence on climate policies. Together they can be seen as an advocacy coalition, sharing the concern for climate change consequences in China and requesting that climate change considerations be reflected in China's development strategy (Stensdal, 2014, pp. 114, 130).

Supply–demand relationship

The supply side of climate change policies is larger than the demand for such policies in China; the turn towards a lower-carbon development is governmentally driven. The central government supplies policies contributing to GHG mitigation out of self-interest: reducing China's anthropogenic GHG emissions, securing continued economic growth, energy supply and air pollution alleviation. The government's policy options are affected by the constraints posed by China's material parameters: there are limited energy resources, but the most polluting option, coal, is cheap and widely available. In providing for the world's largest population, the government receives two seemingly contradictory demands from society. First, to continue the economic growth that moves China towards being a well-off society, which requires more energy sources and until now has been fuelled mainly by coal. Second, there is also a recent and explicit demand for clean air, which means that coal consumption must be reduced. The institutional and political parameters show that despite being a unitary one-party state, the policy-supplying government is not unified, although it outwardly keeps up appearances. NDRC is the most powerful entity under the top leadership. Societal demand is not unimportant, however, as the air pollution example shows. One concerned citizen might not be heard alone, but when enough people express dissatisfaction publicly, particularly enabled by the internet, public opinion is a force to be reckoned with. Also, local governments have taken policy actions to reduce air pollution. Twelve of the provinces, which together consume 44 per cent of the coal consumed in China, have adopted measures to limit their coal use. In terms of absolute reductions, Beijing aims for 50 per cent, Chongqing 21 per cent and Shandong – burning as much coal as Germany and Japan together – aims for a 5 per cent reduction, all in the time frame of 2013 to 2017 (Li and Myllyvirta, 2014). In this case the mentioned local governments push the policy change forward.

For the government, showing leadership on air pollution is important to accommodate public demand. The public was not calling for climate change policies, but rather for air pollution reduction, which shows that climate policy change can be a co-benefit from other policy issues. A similar societal demand for energy policies purely to combat climate change is less likely in a five-year perspective. As in most countries, Chinese citizens care most deeply about their own well-being. The clearest proponents of climate change policies are researchers at select institutions and ENGOs. Through their cooperation with and support of the government they have been somewhat successful in driving policy change. Local governments and businesses operate in the context of the nationally designated policy priorities. They can act as veto players at the implementation stage. The pilots illustrate that the local governments with the delegated responsibility of implementing policies do not always have the needed power to enforce that companies follow their designated policies. By delaying compliance the businesses expressed dissatisfaction with the new policy. A centrally adopted policy is not closed for further negotiation at the implementation stage. On the other hand, ETS is a central government priority and also a project with prestige tied to it for the pilot governments. Therefore, the companies' refusal to comply only delayed implementation of the ETS rather than stopped it.

LIKELY POLICY TRAJECTORY TOWARDS 2020

China is slowing down its emissions increase and controlling its total energy consumption. Given the practice of Five-Year Plans, China's key policy trajectory is fairly predictable. What is clear from the evidence is that energy and climate policies are adopted and successfully implemented when they also achieve other benefits; a policy with win-win options is more likely to be successfully followed through than is a policy that 'costs' only the stakeholders. Despite efforts to limit their growth, China's GHG emissions will continue to increase until 2020 and beyond, but the government expects emissions to peak around 2030.

Given the importance attached to the carbon intensity target, one can expect that the top leadership will impel the provinces to work towards reaching at least a 40 per cent reduction in carbon intensity by 2020 compared to 2005 levels. In about 2020 a national ETS will come into place, if not with full coverage by then. To support the ETS, a cap or limit either on the county's energy use or on emissions is necessary for the effectiveness of the carbon market. China aims to control its energy consumption at 4.8 billion tons of standard coal equivalents, which translates to an annual energy consumption increase no higher than 3.5 per cent until 2020 (Government of China, 2014c).

The energy mix is changing, improving China's energy intensity. China's demand for natural gas will increase, from under 6 per cent currently to at least 10 per cent of primary energy by 2020. Likewise, the installation of non-fossil capacity will continue, and reach 15 per cent of the energy mix by 2020. Coal's share is prescribed at 62 per cent or below in 2020. The government has also made it a goal for China to be 85 per cent self-sufficient in energy in 2020. To this end, energy efficiency measures are crucial. The government also has ambitious plans to develop domestic energy resources such as gas, both conventional and unconventional. Import of the remaining 15 per cent of the energy needed implies that China will be a large actor on the world energy markets.

There is political will in the central government to steer China's future development towards a lower-carbon economy. There are several pros supporting such a decision: diversifying the energy supply, creating new industries, achieving less local pollution and providing more modern buildings and infrastructure. The popularity of air pollution as a topic covered in media and discussions online might come and go in waves. Once the public's awareness of the problem has been established, however, the societal demand to cope with it is not likely to wane until air quality has been significantly improved. Unfortunately, that will likely be after 2020. This societal force might be just as effective in spurring local governments to implement the transition to lower-carbon energy use as the centrally decided targets and goals are. Managing energy needs, economic growth, environmental concerns and policy implementation supervision, among many voices arguing for their interests without a macro outlook, every day is a winding road for the central leadership.

NOTES

1. 'PM2.5' is a term that refers to airborne particulate matter with a diameter no larger than 2.5 micrometres. They are considered dangerous because they can penetrate the lungs and cause lung damage.
2. PM10 is airborne particulate matter consisting of particles with a diameter between 2.5 and 10 micrometres.
3. Criticism is not necessarily censored; the function of Weibo censorship is largely to prevent collective actions (King et al., 2013, pp. 333–40).

REFERENCES

Andrews, S.Q. (2008), 'Inconsistencies in air quality metrics: "blue sky" days and PM10 concentrations in Beijing', *Environmental Research Letters*, **3**, 034009.

Andrews-Speed, P. (2012), *The Governance of Energy in China. Transition to a Low-Carbon Economy*, Basingstoke, UK: Palgrave Macmillan.

Bian, Y., P. He and H. Xu (2013), 'Estimation of potential energy saving and carbon dioxide emission reduction in China based on an extended non-radial DEA approach', *Energy Policy*, **63**, 962–71.

Cao, L., M. Gemmer and T. Jiang (2012), 'Adaptation to climate change in China. Policy, action, and progress', in W. Wang and G. Zheng (eds), *China's Climate Change Policies*, London: Routledge, pp. 58–75.

CCTV News (2013), 'Minister takes lead in promoting electric vehicles', *CCTV.com*, 7 March 2013, accessed 10 May 2014 at http://english.cntv.cn/20130307/107788.shtml.

Chen, G. (2012), *China's Climate Policy*, London: Routledge.

Chen, K. and S. Reklev (2014a), 'Beijing emitters ignore carbon scheme, question govt authority: media', *Reuters*, 13 June, accessed 24 April 2015 at http://www.reuters.com/article/2014/06/13/us-china-carbon-idUSKBN0EO0MR20140613.

Chen, K. and S. Reklev (2014b), 'China's Guangdong extends carbon deadline, plans new auction', *Reuters*, 12 June, accessed 24 April 2015 at http://uk.reuters.com/article/2014/06/12/china-carbon-idUKL4N0OT13S20140612.

Chen, K. and S. Reklev (2014c), 'Guangdong carbon scheme sparks China green vs growth dispute', *Reuters*, 13 May, accessed 24 April 2015 at http://www.reuters.com/article/2014/05/12/us-china-carbon-idUSBREA4B0V020140512.

Chen, K. and S. Reklev (2014d), 'Qingdao city next in line to set up emissions market in China', *Reuters*, 18 February, accessed 24 April 2015 at http://uk.reuters.com/article/2014/02/18/china-carbon-idUKL3N0LN2GI20140218.

China Daily (2014), 'China, US unveil ambitious climate change goals', *China Daily.com*, 12 November 2014, accessed 15 December 2014 at www.chinadaily.com.cn/china/2014-11/12/content_18902550.htm.

Climate Group (2011), *Prospects for Carbon Trading in China*, Beijing: Climate Group.

Donald, S.H. and R. Benewick (2005), *The State of China Atlas*, Berkeley, CA: University of California Press.

Downs, E. (2007), 'China's energy bureaucracy: the challenge of getting the institutions right', in M. Meidan (ed.), *Shaping China's Energy Security. The Inside Perspective*, Paris: Asia Centre, pp. 64–89.

Government of China (2014a), 务院办公厅关于调整国家应对气候变化及节能减排工作领导小组组成人员的通知 [State Council Office's notification on the adjustment of personnel in the national leading working group for climate change response and energy saving and emission reduction], accessed 10 June 2014 at www.gov.cn/zwgk/2013-07/09/content_2443020.htm.

Government of China (2014b), 国务院办公厅关于调整国家能源委员会组成人员的通知 [State Council's notification on the adjustment of the composition of the National Energy Commission personnel], accessed 10 June 2014 at www.gov.cn/zhengce/content/2013-07/11/content_7570.htm.

Government of China (2014c), 控总量●调结构●促改革––解读《能源发展战略行动计划（2014–2020年）》三大看点 [Controlling the total amount. Adjusting the structure. Promote reform – interpreting the Strategic Action Plan for Energy Development (2014–2020). Three important points], accessed 1 December 2014 at www.gov.cn/xinwen/2014-11/19/content_2780851.htm.

Hallding, K., G. Han and M. Olsson (2009), 'China's climate- and energy-security dilemma: shaping a new path of economic growth', *Journal of Current Chinese Affairs*, **38**(3), 119–34.

Heggelund, G., S. Andresen and I.F. Buan (2010), 'Chinese climate policy: domestic priorities, foreign policy and emerging implementation', in K. Harrison and L.M. Sundstrom (eds), *Global Commons, Domestic Decisions*, Cambridge, MA: MIT Press, pp. 239–61.

International Energy Agency (IEA) (2014a), 'China (People's Republic of China and Hong Kong China): Balances for 2012', accessed 10 December 2014 at www.iea.org/statistics/statisticssearch/report/?&country=CHINAREG&year=2011&product=Balances.

International Energy Agency (IEA) (2014b), *Key World Energy Statistics 2014*, Paris: IEA.

King, G., J. Pan and M.E. Roberts (2013), 'How censorship in China allows government criticism but silences collective expression', *American Political Science Review*, **107**(2), 326–43.

Kostka, G. and W. Hobbs (2012), 'Local energy efficiency policy implementation in China: bridging the gap between national priorities and local interests', *The China Quarterly*, **211**, 765–85.

Lampton, D.M. (1992), 'A plum for a peach: bargaining, interest, and bureaucratic politics in China', in K.G. Lieberthal and D.M. Lampton (eds), *Bureaucracy, Politics, and Decision Making in Post-Mao China*, Berkeley, CA: University of California Press, pp. 34–59.

Li, J. (2014), '670,000 smog-related deaths a year: the cost of China's reliance on coal', *South China Morning Post*, 5 November, accessed 6 November 2014 at www.scmp.com/news/china/article/1632163/670000-deaths-year-cost-chinas-reliance-coal.

Li, S. and L. Myllyvirta (2014), *The End of China's Coal Boom. Six Facts You Should Know*, Beijing: Greenpeace.

Lieberthal, K.G. (1992), 'Introduction: the "fragmented authoritarianism" model and its limitations', in K.G. Lieberthal and D.M. Lampton (eds), *Bureaucracy, Politics, and Decision Making in Post-Mao China*, Berkeley, CA: University of California Press, pp. 1–31.

Martina, M. (2013), 'China premier-designate says pollution solution "long-term process"', *Reuters*, 15 January, accessed 16 May 2014 at http://uk.reuters.com/article/2013/01/15/china-pollution-idUKL4N0AK5E920130115.

MEP (2013), 'The State Council issues Action Plan on Prevention and Control of Air Pollution introducing ten measures to improve air quality', accessed 15 May 2014 at http://english.mep.gov.cn/News_service/infocus/201309/t20130924_260707.htm#.

National Bureau of Statistics of China (2014), *China Statistical Yearbook 2013*, Beijing: China Statistics Press.

National Development and Reform Commission (NDRC) (2013), *China's Policies and Actions for Addressing Climate Change*, Beijing: NDRC.

National Development and Reform Commission (NDRC) (2014a), 国家应对气候变化规划(2014–2020)年 [National Plan for Addressing Climate Change (2014–2020)], Beijing: NDRC.

National Development and Reform Commission (NDRC) (2014b), 能源发展战略行动计划 (2014–2020 年) [Strategic Action Plan for Energy Development (2014–2020)], accessed 10 December 2014 at www.gov.cn/zhengce/content/2014-11/19/content_9222.htm#.

Pew Research Center (2013), 'Environmental concerns on the rise in China. Many also worried about inflation, inequality, corruption', survey report, Washington, DC: Pew Research Center.

Saich, T. (2011), *Governance and Politics of China*, 3rd edition, Basingstoke, UK: Palgrave Macmillan.
Spegele, B. (2012), 'Comparing pollution data: Beijing vs. US Embassy on PM2.5', *Wall Street Journal*, 23 January, accessed 16 May 2014 at http://blogs.wsj.com/chinarealtime/2012/01/23/comparing-pollution-data-beijing-vs-u-s-embassy-on-pm2-5/tab/print/.
Stensdal, I. (2014), 'Chinese climate-change policy, 1988–2013: moving on up', *Asian Perspective*, **38**(1), 111–35.
Sydney Morning Herald (2014), 'China aims high for carbon market by 2020', *Environment*, 12 September, accessed 24 April 2015 at http://www.smh.com.au/environment/climate-change/china-aims-high-for-carbon-market-by-2020-20140911-10ftdp.html.
Wee, S.-L. (2014), 'Chinese man becomes first to sue government over severe smog', *Reuters*, 25 February, accessed 13 May 2014 at http://in.reuters.com/article/2014/02/25/us-china-pollution-idINBREA1J18E20140225.
World Bank (2014), 'World DataBank', accessed 5 November 2014 at http://databank.worldbank.org/data/home.aspx.
Wübbeke, J. (2013), 'China's climate change expert community – principles, mechanisms and influence', *Journal of Contemporary China*, **82**(22), 712–31.
Xinhua (2013), 'PM2.5 in China's air quality standards', accessed 16 May 2014 at www.china.org.cn/environment/2012-03/01/content_24767087.htm.
Ye, Q., L. Ma, H. Zhang and H. Li (2008), 'Translating a global issue into local priority: China's local government response to climate change', *The Journal of Environment & Development*, **17**(4), 379–400.

4. EU climate and energy policy: demanded or supplied?

Jon Birger Skjærseth*

BOX 4.1 KEY FACTS ABOUT THE EUROPEAN UNION

Population 2014	508 million
Population growth rate 2013	0.2%
Population density 2013	119.6 per km^2
GDP per capita 2013	35 438 USD
GDP growth rate 2013	0.1%
GINI index 2013[a]	30.5
Access to electricity 2010	100%
Land area 2014	4 236 562 km^2
Political system	The EU is a complex and decentralized multi-level political system based on the voluntary commitment of its member states. The basic EU institutions are the Commission, the Council, the European Parliament and the European Court of Justice
Subnational entities	28 member states

Note: a. Eurostat (2014).
Source: World Bank (2014).

INTRODUCTION

Since the 1990s, the European Union (Box 4.1) has sought to play a leadership-by-example role in international climate negotiations, adding increasingly ambitious targets and policy instruments. In October 2014, EU leaders agreed to reduce greenhouse gas (GHG) emissions within the EU by at least 40 per cent from 1990 levels by 2030. This target is in line with an EU trajectory towards at least an 80 per cent GHG reduction by

2050. In 2007, the European Union adopted targets for cutting GHG emissions, increasing the share of renewable energy consumption and energy efficiency by 20 per cent by 2020 compared to 1990 levels. These targets were followed by a package of binding climate and energy policies. EU GHG emissions have declined and consumption of renewable energy has increased alongside energy saving.

Its climate policy makes the European Union a leader in the international arena, but this can hardly explain why the European Union seems to be so far ahead of other major emitters. How has the European Union been able to adopt a relatively ambitious climate policy based on agreement between 28 member states with significantly different energy economic situations?

This chapter takes as its point of departure the 'supply–demand' policy approach of this volume (see Chapter 1). It deviates from traditional explanations of EU integration and policy-making that have been based on approaches such as liberal intergovernmentalism and multilevel governance (see below). It also deviates somewhat from the supply–demand policy approach, because the EU political system diverges in important ways from other political systems – regarding how governments supply policies, how society demands policies and how these demands are channelled. This divergence is reflected in my analysis, which pays most attention to the policy supply side structured around policy phases: initiation, negotiation, implementation and reform towards 2030.[1]

The next section contains an analysis of the development of EU climate and related energy policies. Then I explain this development from the perspective of energy security needs and varying actor interests, the institutional decision-making machinery and related ability to link policies and issues to promote agreement and implementation, and public demand for climate and energy policies as expressed in Eurobarometer polls. In the subsequent sections, the future for EU climate policy is explored, and then conclusions are presented.

DEVELOPMENT OF EU CLIMATE POLICY

The story of how EU climate policy has evolved is about how climate *and* energy policies have been combined. In 1991, the European Union unsuccessfully sought to adopt a package of climate and energy policies based on carbon/energy pricing, renewables and energy efficiency. The package was intended to show leadership-by-example before the adoption of the United Nations Framework Convention on Climate Change (UNFCCC) at the Rio Conference in 1992 (Skjærseth, 1994). Thereafter,

EU climate and energy policies developed slowly and separately, but both policy strands gained momentum with the 1997 Kyoto Protocol.

Table 4.1 GHG emissions and energy data (EU27)[a]

Total GHG emissions by year	1990	2005	2011
	5574 $MtCO_2e$ [b,c]	5129 $MtCO_2e$	4550 $MtCO_2e$
GHG emissions per capita 2012			8.95 tCO_2e
GHG emissions by sector 2012	Electricity and heating 27%, transport 19.7%, manufacturing industries and construction 11.7%, agriculture 10.3%, residential 9.4%, industrial processes 7.1%[d]		
Energy mix 2011 (gross inland consumption)	Fossil fuels 76%, nuclear 14%, renewables 10%		

Notes:

a. Eurostat data (2014).
b. Million metric tons CO_2 equivalents.
c. Excluding land use, land use change and forestry (LULUCF); see EEA (2013).
d. EEA (2014a).

In 2001, a directive on promoting electricity from renewable energy sources was adopted (2001/77/EC), followed in 2003 by a directive on substituting biofuel for diesel and petrol in transport (2003/30/EC). These directives aimed primarily at promoting security of energy supply, and secondarily at implementing the EU's 8 per cent GHG reduction commitment under the Kyoto Protocol. Policies on energy efficiency followed the same priority in objectives (Commission, 2005). However, energy policy targets were non-binding, reflecting the common view that member states should have sovereignty in choosing their primary energy sources and their energy mix.

In October 2001, the European Commission launched a climate policy proposal for an EU emissions trading system (ETS), intended as an innovative market-based policy instrument with its first ever international cap-and-trade system targeting 11 000 industrial installations. The final ETS Directive was adopted unanimously by the 15 member states in late 2002, approved by the European Parliament, and formally adopted in July 2003. A first pilot phase ran from 2005 to 2007, followed by the Kyoto commitment phase, 2008–12. In 2004, the European Union adopted the associated Linking Directive, connecting the ETS to the Kyoto Protocol's flexible project mechanisms, mainly the Clean Development Mechanism (CDM).[2] The ETS was used by the EU as a leadership-by-example strategy to persuade enough states to ratify the Kyoto Protocol for it to enter into force (Skjærseth and Wettestad, 2008).

In March 2007, the heads of states and governments in the European Council unanimously adopted the 20-20-20 targets: to cut GHG emissions, to increase the share of renewables, and to increase energy efficiency by 20 per cent by 2020 compared to 1990 levels (European Council, 2007, p. 12). In January 2008, the Commission formally proposed the climate and energy package for achieving the targets by 2020. The main structure involved two cross-sector instruments. The first was a revised EU ETS aimed at reducing emissions in the ETS sectors by 21 per cent below 2005 emission levels. The proposal included a transition from a decentralized system to an EU-wide cap to be reduced annually by 1.74 per cent. Allocation procedures were altered and harmonized, from free allowances to a system based on payment-by-auctioning as the main principle. The second instrument was an effort-sharing decision (ESD) based on different national targets, to yield a 10 per cent reduction for sectors not covered by the ETS, such as transport, agriculture, waste and buildings. In addition, the core package contained two technology-specific directives: one on the promotion of renewable energy sources (RES) based on different national targets, and a proposal for a legal framework for safe carbon capture and storage (CCS). Other transport policies simultaneously developed by the Commission were prepared according to a different time schedule.[3]

The package's main structure remained intact throughout the 2008 negotiations, despite several changes to the Commission's original proposal (Skjærseth, 2013). The package, formally adopted in 2009, was to be implemented with a view towards achieving a low-carbon economy by 2050. In October 2009, the European Council agreed to support an EU goal of reducing GHG emissions between 80 and 95 per cent by 2050 compared to 1990 levels (European Council, 2009).

Alongside implementation of the package, the EU continued to develop new long-term climate policies. In 2011, the Commission published two roadmaps for moving towards a low-carbon economy by 2050. Both were supported by 26 of the 27 member states, but vetoed by Poland. Despite these setbacks, in January 2014 the Commission formally proposed the new climate and energy policy framework for 2030, finally adopted in modified form by the European Council in October (Commission, 2014c). The new goals are domestic reductions of at least 40 per cent GHG emissions, a 27 per cent increase in renewable energy consumption at EU level, and an indicative target of a 27 per cent increase in energy efficiency – all by 2030 compared to 1990 levels. Part of the agreement involves strengthening the EU's ETS by lowering the cap from 2021.[4] Emissions from non-ETS sectors are to be cut by 30 per cent below the 2005 level, through continued effort sharing among the member states.[5]

Thus, EU climate policy has grown more ambitious in its goals and policy instruments. This growth has strengthened the credibility of the EU's ambitions of being a leader-by-example in the international climate negotiations.

EXPLAINING EU CLIMATE POLICY

EU climate policy has been explained through approaches like liberal intergovernmentalism, multilevel governance and international regimes (Skjærseth and Wettestad, 2008; Skjærseth, 2013). Liberal intergovernmentalism places main emphasis on the member state level; multilevel governance on the EU supranational and non-state levels (in addition to member states); and the regime approach on institutional factors and processes external to the European Union. In this chapter, which has been structured to fit a comparative study of key actors in climate policy, these approaches serve mainly as a context for understanding the following: energy security concerns and related actor interests; institutional problem-solving capacity; and how policies and issues have been linked to initiate, decide, implement and reform EU climate and energy policies. Supply of policies will in turn be related to societal demands, for better explaining the principal drivers behind EU climate policy.[6]

Energy Security

In 2012, EU energy demand was similar to 1995 levels. Because primary energy production decreased by almost one-fifth in the same period (Commission, 2014a, 2014b), EU energy import dependency has been on the rise since the mid-1990s. In 2012, 53 per cent of EU energy consumption was linked to imports.

Energy imports to the European Union cost about one billion euros per day and energy dependency is 88 per cent of crude oil, 66 per cent of natural gas, 42 per cent of coal and 95 per cent of uranium (Commission, 2014b). For oil and coal imports, energy security is largely a question of affordability.[7] Uranium for the EU's 131 nuclear power plants in 16 member states can be purchased from multiple suppliers, but the final fuel assembly process is managed by one Russian company for member states with Russian-designed reactors. Gas supplied through pipelines is the major concern. Several member states are heavily reliant on Russian gas: the Baltic republics, Slovakia, Bulgaria and Finland all depend solely on Russia's Gazprom. Moreover, Estonia, Latvia and Lithuania rely on one Russian operator for electricity operation, network and a large share of supply.

Energy security is mainly linked to the development of EU climate and energy policies by driving energy efficiency, by increasing energy production, by completing the internal gas and electricity market, and by improving infrastructure to provide greater flexibility in directing energy across the European Union to where it is needed. Energy security concerns have climbed the EU political agenda concurrent with gas shortages through Ukraine transit routes in 2006, 2009, and now in the crisis in Ukraine.[8]

Still, interests and opinions vary among the member states on the extent to which a transition to a low-carbon economy will promote energy security. While climate change and energy security concerns are seen as synergetic by countries such as Germany, they are seen as conflictual by some energy-dependent Central and East European countries (CEECs) with indigenous fossil fuels. From a Polish governmental perspective, for example, an ambitious climate policy based on carbon pricing and renewable subsidies will undermine coal, threaten physical energy security and lead to less affordable electricity (Skjærseth, 2014a).[9]

In other words, there is a long way from energy security concerns to climate action agreement because of the differing interests among member states.

Institutions for Change

The European Union is not a state with exclusive rights to the legitimate use of coercion; rather, it is a complex and decentralized multilevel political system based on the voluntary commitment of its member states. It relies on these states to administer state power, but it can work as a complete political system without transforming the territorial organization of the state (Hix, 2005). The European Union also departs from a classic federal state such as the United States in important ways, including separation of powers, cooperative norms and capacity for long-term policy-making (Skjærseth et al., 2013).

The major EU institutions are the Commission, the Council, the European Parliament and the European Court of Justice.[10] The Commission has various responsibilities, including development of policy ideas, initiating legislation and scrutinizing implementation. The over 30 Directorates General (DGs) differ in interests and culture, and tend to put forward competing policy positions – not only in climate and energy policy (Hix, 2005). While compartmentalization may prevent policy initiation, a distinctive strength of the Commission is its ability to shape and frame (climate) policies for the long term. The non-partisan Commission, led by commissioners from each member state, is collectively responsible for decisions, and is appointed to serve EU interests. This enables it to think in a more long-term manner than most individual member states can (Skjærseth et al., forthcoming). Nevertheless, commissioners are also a link to national publics and serve as a crucial clearinghouse for the interests of the member states vis-à-vis the Commission (Egenhofer et al., 2011).

The Council of the European Union consists of representatives at ministerial level from the 28 member states. Its main function is to pass EU laws mainly by qualified majority; here it shares the final say with the European Parliament (EP).[11] The EP is directly elected every five years. It shares decision-making authority with the Council and is generally seen as the 'greenest' of the EU institutions. Power dispersion among EU institutions leads to extensive consultations among the Commission, Council and EP before formal negotiations on legislation, a practice that encourages consensus seeking in decision-making.

The European Council, where the heads of states and governments meet normally four time a year, has become increasingly important in developing climate and energy policy. The EU leaders adopted the 20-20-20 targets for 2020 and the new climate and energy policy framework for 2030. The European Council does not exercise legislative functions, but defines the political direction and priorities of the European Union. Decisions are normally taken by consensus.

Since the Treaty of Rome in 1957, successive treaties and treaty reforms have given these institutions more legislative, executive and judicial powers, culminating with the 2003 Nice Treaty and the 2009 Treaty of Lisbon.[12] The Lisbon Treaty introduced a number of changes, including granting more power to the EP and establishing more policy areas subject to qualified majority voting. New energy provisions in the Lisbon Treaty mainly reflect codification of existing practices (see Braun, 2011).

As a result of the growing institutional powers of EU institutions, citizens and an increasing number of groups are seeking to affect EU policy-making.[13] The over 0.5 billion citizens in the 28 EU member states, who voice their demands through various channels, are the basis for the EU political system (Hix, 2005). First, they elect national parliaments that form national governments that are represented in the Council and the European Council. Second, citizens elect Members of the European Parliament (MEPs) in EU-wide elections.[14] Third, they can take legal action in national courts and the European Court of Justice (EJC). Finally, citizens can form interest groups that seek to exert influence on EU legislation through a variety of channels, not least through lobbying national governments and EU institutions. Interest groups, such as environmental NGOs, large companies and European industry associations, have played important roles in shaping and implementing EU climate and energy policies. Their participation became institutionalized from March 2000, when the Commission launched the European Climate Change Programme (ECCP), intended to enable stakeholders to participate in preparatory work on new climate policies. Stakeholders here included various Commission DGs, member state experts, industry and other non-governmental organizations that tend to form policy 'networks' around specific interests and policies.

The 'feedback' between supply of and demand for EU policies is relatively weak compared to that in other political systems (Hix, 2005). National elites are the main door-keepers of EU information, which is filtered through national media that focus on national rather than EU-level issues.

We can conclude that the EU institutional setting promotes consensus seeking and long-term policy development. The EU political system is extremely complex and deviates in important ways from other political systems in how governments supply policies, in how society demands policies, and in how these demands are channelled. This will be reflected in the following analysis, where supply and demand will be integrated in policy phases.

Supply and Demand of EU Climate Policies

Initiation

Because EU climate and energy policies were adopted by consensus from 2007, the explanations for change are mainly related to how policies and issues were combined in new ways that shared the costs, promoted new low-carbon opportunities and gave something to all the major 'veto players' involved in policy-making (Tsebelis, 2002).[15]

By 2006, energy security climbed the political agenda also because of rising oil prices and the Ukraine–Russia energy dispute that threatened European gas supplies. In March, the Commission issued a Green Paper, prepared by DG Energy, on a European strategy for sustainable, competitive and secure energy (Commission, 2006). It painted a dark picture of energy challenges and stressed the need for a new energy policy as an *integrated* part of EU climate policy by framing climate and energy policies as complementary and synergistic.

A handful of individuals with close ties to the Environment and Energy DGs and to the top level of the Commission were able to make this happen (Skjærseth, 2013). The three biggest EU member states – Germany, France and the United Kingdom – were on board as well. The United Kingdom had also shifted from resisting to supporting EU-level energy policies during its autumn 2005 presidency (Eikeland, 2012).

With 2007 came a turning point for EU climate and energy policies. In January, the Commission issued two key communications on energy and climate policy strategies for 2020 and beyond (Commission, 2007a, 2007b). These communications set a radical tone: 'Strong scientific evidence shows that urgent action to tackle climate change is imperative' (Commission, 2007a, p. 3). The response was to transform Europe, catalysing a new industrial revolution by speeding up the change to low-carbon growth: 'The EU would have set the pace for a new global industrial revolution' (Commission, 2007b, pp. 5, 21). These communications proposed the 20-20-20 targets. The target of reducing GHGs by at least 20 per cent would be stepped up to 30 per cent if an adequate international climate treaty could be agreed. The twin targets represented an effort to demonstrate international leadership on a post-2012 climate treaty (Commission, 2007a). The main new legislation signalled was on the ETS, renewables and CCS.

Synergies between climate and energy were underscored. Action on climate change was placed at the centre of a new EU energy policy by stressing the importance of: making energy use more efficient, lessening the need for imported hydrocarbons and reducing vulnerability to fluctuations in oil and gas prices. Action on energy policy would contribute to climate change mitigation and more effective application of the ETS, while also creating new 'green' jobs. These were to be achieved by strengthening policies on renewables, energy efficiency, liberalization of the European energy market and technological innovation. A European Strategic Energy Technology Plan (SET-Plan) was proposed, to lower the cost of clean energy and put the European Union at the forefront of the low-carbon technology sector.[16]

The proposals also rested on linking differently valued issues, particularly among member states. The ten CEECs that joined the EU between 2004 and 2007 were more concerned about energy security, whereas most of the EU-15 favoured a more stringent climate policy, as expressed by the adoption of the ETS. The CEECs were less prosperous and less energy efficient than the EU-15, increasing interregional EU difference in carbon intensity and GDP per capita. Moreover, many CEECs relied heavily on coal, which could have a prosperous future with CCS.

The EP responded swiftly, in February 2007 issuing a resolution on climate change that argued for targets even more stringent than those proposed by the Commission (European Parliament, 2007). Green groups argued that the 20 per cent GHG target was insufficient and demanded a 30 per cent unilateral reduction goal. Energy-intensive industry expressed concerns about competitiveness if the European Union committed ahead of other major emitting countries. The renewables industry, represented by the European Renewable Energy Council (EREC), supported stronger action on renewables, but opposed replacing sector targets (e.g., on electricity) with an overall renewables target (EREC, 2007).

In March 2007, the European Council adopted the key elements of the new integrated climate and energy policy, noting the central role of the EU ETS in the EU's long-term strategy for reducing GHG emissions (European Council, 2007, p. 12). In addition to new measures for improving the internal market and security of supply, the Council adopted the new targets on energy efficiency, renewable energies and energy technologies. With regard to technology, the Council welcomed the construction and operation by 2015 of up to 12 demonstration plants for CCS. The renewables target and support mechanism were particularly controversial (Ydersbond, 2014). Many of the coal-reliant new members, with Poland in the lead, were concerned that the RES targets could force them to invest in sources of renewable energy that were more expensive. Despite these warnings, the Commission had gained backing from the EU leaders and the EP – the two main veto players – before the package entered the formal decision-making phase.

The initiation of new climate and energy policies only partly reflected societal demands as expressed in Eurobarometer polls. A spring 2006 survey showed that the national level was perceived as most appropriate for energy policies (Eurobarometer, 2006), especially in the new member states. Moreover, the majority of Europeans (59 per cent) were not prepared to pay more for energy produced by renewable sources. Although half of those surveyed were willing to reduce their energy consumption, only 4 per cent would make this change if it meant paying more. Thus, there was scant public demand for strengthening and harmonizing policies on renewables and energy efficiency at the EU level when the climate and energy package was initiated.

EU-level climate policies enjoyed strong public support. By 2006, climate 'hype' was sweeping Europe. Attention to climate change was also raised by Al Gore's award-winning documentary, *An Inconvenient Truth*, released in May 2006, and by warnings in the October *Stern Review* that failure to act on global warming would impose major costs on the global economy. Support for EU climate action was confirmed by a wave of Eurobarometer surveys. Climate change had clearly become a major concern, with three-quarters of those surveyed confirming that they took the issue very seriously (Eurobarometer, 2008). They felt that industry, citizens, national governments and the European Union were not doing enough to fight climate change.[17] And, asked about the 20-20-20 targets, a majority considered the targets to be set at levels that were 'about right' or 'too modest'.

Negotiation

In January 2008, the Commission formally proposed legislation that would make up the climate and energy package of binding policy instruments. Burden, or effort sharing, had gained a more prominent place since 2007. Overall costs to EU economies by 2020 were estimated at just under 0.5 per cent of GDP. In principle, no member state was expected to undertake investments that diverged too sharply from this average. The EU climate and energy package was based on a thorough assessment of how the ETS, ESD and RES proposals would work together to level the costs (Commission, 2008). Significant differences in costs by 2020 between member states were levelled out to compensate poorer ones: by setting different national targets in the non-ETS sectors (ESD), based on GDP/capita; by setting different national targets for the share of EU energy consumption to be achieved by renewable energy (RES), based on a combination of GDP and a flat-rate increase in the share of renewable energy; by using auctioning revenues (ETS) to compensate lower-income member states. Analyses of energy-intensive industries exposed to significant international competition and at risk of carbon leakage had showed that granting access to cheaper CDM credits and providing free allowances through benchmarking would be effective strategies for limiting potentially negative effects on competitiveness under the revised ETS.

Nothing guarantees that the Commission's legislative proposals will survive negotiations.[18] Still, the package's main structure remained intact throughout the 2008 negotiations. Member states unanimously adopted the package and the EP endorsed it, and endorsed the reduction targets for ETS and non-ETS sectors and different national targets in the ESD and RES Directives. The four legislative proposals were complex, and the schedule for their development was tight. The EU needed an ambitious package to show its 'leadership-by-example' before the international climate negotiations in Copenhagen and the June 2009 EP elections. Deliberations commenced with informal negotiations among high-level representatives from the Commission, the EP and the Council. The major climate policy element in the package was the proposal for revising the EU ETS. Here there were two main areas of disagreement: first, the new CEECs, led by Poland, demanded more economic 'solidarity' than originally included in the Commission's proposal and demanded free allowances for the electric power industry; and second, energy-intensive industries demanded more free allowances, to reduce the risk of their pulling out of Europe (Skjærseth and Wettestad, 2010). These demands were voiced with increasing intensity in autumn 2008, fuelled by the unfolding global economic crisis and rising concerns among policy-makers about the costs of the package. However, public opinion on climate change and climate policies was apparently not affected; a Eurobarometer survey in 2009 confirmed that previous trends had prevailed. Moreover, about 60 per cent believed that fighting climate change and protecting the environment could have positive impacts on the economy, by creating green jobs and greener industries (Eurobarometer, 2009).

In autumn 2008, France took over the EU Presidency from Slovenia; French leadership proved extremely important in forging compromise by year's end (Skjærseth and Wettestad, 2010). One institutional tool used was the introduction of a more demanding decision-making procedure, replacing qualified majority voting in the Council of Ministers with unanimity in the European Council. Because the package's elements were mutually reinforcing, decision-makers were pressured to negotiate and adopt all legislative proposals simultaneously: in practice, each member state plus the EP would have a veto position. The comprehensive package designed by the Commission provided ample room for exchanging concessions across different policies. For example, Germany defended energy-intensive industries and demanded more free allowances than proposed by the Commission, but the EP opposed this. In return for accepting more free allowances, the EP got 300 million allowances from the ETS New Entrants' Reserve (NER 300) for co-financing up to 12 CCS demonstration projects and renewable technologies, linking the new CCS legislation to the ETS. But everyone was not equally happy with the outcome (Skjærseth, 2014a). Poland had threatened to veto the whole package. It strongly opposed a more ambitious ETS and managed to get several concessions on solidarity and free allowances for the electric power sector – but these concessions were not enough to meet Polish demands. On RES, Poland was positive to co-firing biomass and coal in existing coal burners, but had to accept a somewhat stricter RES target than it preferred. Poland welcomed the CCS Directive and the NER 300. The proposed Polish ESD target of 14 per cent increase in emissions was seen as generous and was accepted without resistance.

Implementation

Implementation of the package from 2009 took place concurrently with the development of new long-term climate policies. The challenge at the national level was to reproduce synergies and compensation that had promoted EU-level agreement. In Poland, the combination of climate and energy policies did not promote implementation or support for long-term EU decarbonization policies. Climate change policies have few supporters in Poland whereas energy security is seen as important for society and by policy-makers. Policy-makers view an ambitious EU climate policy as a potential threat to energy security in that it may threaten coal, shale gas exploration or affordable electricity prices (Skjærseth, 2014a).

Subsidies to promote RES in Poland have only to a limited extent spurred 'green' growth in terms of new jobs resulting from technology-neutral green certificates and co-firing biomass and coal as the preferred and cheapest option. CCS might have become a new area for technological innovation and first mover advantages in Poland, but all pilot projects have been cancelled, partly because of the low carbon price and limited NER 300 funding.

As a result of mainly the economic crisis and high imports of external credits, a surplus of ETS allowances had accumulated by early 2012, rising to 2.1 billion by the end of 2013 (Commission, 2012).[19] The carbon price plunged from nearly EUR 30 in spring 2008 to just above EUR 5 in spring 2014, with a significant reduction in the second half of 2011 coinciding with an accelerated build-up of allowances and international credits.

Concerns about air pollution can raise willingness to develop climate policy, particularly in coal-dependent countries. In Poland, however, air pollution is dealt with by end-of-pipe technologies, like application of flue-gas desulphurization facilities. Such policies will have limited impact on CO_2 emissions as long as coal is not replaced by low-carbon technologies (Council of Ministers, Poland, 2008, p. 35). Revenues ranging from auctioning of emission allowances to compensation for low-income member states were effectively used to promote EU-level agreement. In Poland, the low carbon price and exemption from auctioning in the electric power sector (70 per cent from 2013) have prevented this mechanism from having effective impacts.[20] Poland has also opposed measures that could increase the carbon price.[21] Thus, issue linkage mechanisms that promoted EU-level agreement have only to a very limited extent been transformed into new low-carbon opportunities in Poland. Moreover, the EU targets and policies had all been adopted before the 2009 Copenhagen Climate Summit, where there were high expectations that a binding and ambitious climate treaty would be agreed. However, the 'Copenhagen Accord' failed to set binding global reduction targets. Slow progress on a new international climate treaty has contributed to Polish opposition towards new long-term EU climate policies.

The combination of poor experience from implementation and pressure to accept unwanted policies has made Poland increasingly resist new EU climate and energy policies. In March 2011, the Commission published the roadmap prepared by DG Climate Action for moving towards a competitive low-carbon economy by 2050 (Commission, 2011a). The analysis showed that GHG emissions would have to be reduced (from 1990 levels) by 25 per cent by 2020, 40 per cent by 2030 and 60 per cent by 2040 to reach 80 per cent by 2050. In December 2011, the Commission published an Energy Roadmap 2050 prepared by DG Energy (Commission, 2011b). Here the key message is that greater energy efficiency and more use of renewables to achieve 80–95 per cent reduction by 2050 will cost about the same as would continued heavy reliance on nuclear power and fossil fuels. Twenty-six of the 27 EU member states agreed with the presidency's conclusions based on these roadmaps. The blocking state on both occasions was, as noted, Poland. A new Eurobarometer survey from 2011 indicated that the Polish position deviated sharply from general European public opinion, even more so than in 2009 (Eurobarometer, 2011). Fifty-one per cent of respondents saw climate change as one of the world's most serious problems – 20 per cent as *the* most serious problem. Further, 89 per cent perceived climate change as a 'serious problem', with 68 per cent considering it 'very serious'. The share that saw economic benefits in tackling climate change rose to 78 per cent.

Reform

In 2013, the Commission issued a Green Paper on a new framework for climate and energy policies towards 2030 (Commission, 2013). Concerning targets, all member states that responded favoured a GHG reduction target for 2030, except Poland (Ibec, 2013). On renewables, there was disagreement on setting a new target and what form it should take. On energy efficiency, member states also expressed mixed views on the usefulness of a new EU target. There was general agreement that the EU ETS should remain a central instrument.

In January 2014 then, the Commission proposed the revised framework on climate and energy policies for 2030. The proposal reflected the economic crisis, lack of international progress and varying experience with implementation of the climate and energy package. The tone of the proposal had changed markedly from 2007: gone was the rhetoric on a new 'green' industrial revolution, synergies and energy technological innovation. CCS was not mentioned. The proposal included the new target of reducing GHGs domestically by 40 per cent below 1990 levels (Commission, 2014c), to be achieved by increasing the annual ETS cap from the current 1.74 per cent to 2.2 per cent after 2020, and by a market stability reserve for dealing with the accumulating surplus of emission allowances.[22] Targets and policies on renewables were weakened. An EU-wide target of at least 27 per cent renewable energy by 2030 was proposed: this was only slightly above expected developments and would not be translated into new binding national targets.

A new Eurobarometer survey indicated that public demand for EU climate policies had declined somewhat (Eurobarometer, 2013). The number of respondents who saw climate change as the *most serious* problem in the world decreased slightly compared to the 2011 level, but 90 per cent still regarded climate change as a *very serious* problem. About half thought it was 'very important' for their government to support energy efficiency and renewable energy by 2030. Responsibility for dealing with the problem was still placed on industry, national governments and the EU, but 48 per cent said that national governments had the main responsibility, compared to 39 per cent for the EU. This poll also confirmed significant national differences. For example, 81 per cent of Swedish respondents see climate change as one of the most important problems, as against only 28 per cent in Estonia. Respondents in Sweden are also more likely to take action (80 per cent) than are respondents in Estonia (25 per cent) and Romania (23 per cent).

Initial responses to the Commission's proposal show that the member states were deeply divided. On the one hand, prior to the European Council discussion of the framework on 20–21 March, the 'Green

Growth Group' (including Estonia) issued a joint statement.[23] This is an informal grouping of like-minded energy, climate and environment ministers from 13 EU member states who agree on the 2030 framework's core elements as set out by the Commission (Green Growth Group, 2014). On the other hand, with Poland in the lead, Hungary, the Czech Republic, Slovakia, Romania and Bulgaria demanded full national control over their energy mixes, compensation from the EU and burden sharing.[24] Essentially, they held that ambitious member states like Germany, France and the United Kingdom should deliver the reductions towards 2030 themselves. Poland also demanded that new climate and energy policies be made conditional on a global climate deal in Paris 2015.

In October 2014, the European Council adopted the new framework for new EU climate and energy policies for 2030, including the new 40 per cent domestic reduction goal, the 27 per cent renewable energy consumption goal, and an indicative target of a 27 per cent increase in energy efficiency – all compared to 1990 levels. This new framework means that Poland has accepted new long-term EU policies based on the ETS as a principal climate policy instrument *before* the climate negotiations at Paris 2015. However, the framework includes various concessions to Poland and other low-income CEEC member states. First, burden sharing will continue for non-ETS sectors, based on GDP/cap as today. There will be more flexibility in how national targets can be achieved, although the details are not clear. Second, free allocation will continue for energy-intensive industry, to prevent the risk of carbon leakage; future allocation should ensure affordable energy prices and better alignment with changing production levels in different sectors – which may mean that allocation will take increases in production into account.

Third, low-income member states can allocate for free up to 40 per cent of the allowances to the electric power sector, representing a weakening of current policies aimed at gradually reducing free allowances to zero by 2020. Fourth, the 'solidarity funds' of 2 per cent and 10 per cent to modernizing energy systems in low-income member states will continue. Upgrading of coal power is not ruled out by the specifications. Finally, the framework recognizes that energy security can be increased by having recourse to indigenous resources, such as coal and shale gas. Full respect for member-state freedom to determine their energy mix is also explicitly included – as are a non-binding energy efficiency goal and a new RES goal that will not be transformed into binding national goals.

The new EU 2030 framework thus represents a 're-packing' compromise to satisfy the main veto players, with substantial concessions to

Poland and other CEECs. The 2030 framework must be specified by detailed legislation that will determine the ambitiousness of the new policy.

POLICY OUTLOOK

The EU climate and energy package of policies is fixed until 2020. Good progress has been made towards the climate and energy targets (EEA, 2014b). In 2013, GHG emissions were already 19 per cent below 1990 levels, seven years ahead of the 20 per cent target by 2020. The EU is also on track concerning the renewables target: by 2012, the share of RES in gross final energy consumption was 14.1 per cent, ahead of the 13 per cent target for 2012. Moreover, energy consumption is decreasing to meet the 2020 energy efficiency target. However, progress is uneven, and member states display significant variation in experiences with implementing the climate and energy package (Eikeland, 2014; Jevnaker, 2014; Gulbrandsen and Skjærseth, 2014; Skjærseth, 2014a).

In 2007, the European Council endorsed the construction of up to 12 demonstration plants for CCS. Today, there are no new CCS pilot projects left. The carbon market is oversupplied with over two billion ETS allowances that will probably last well beyond 2020. The amount of surplus by 2020 will depend on (in addition to policy reform) the pace of economic recovery in Europe and various energy factors, including the development of renewable energy and energy efficiency in ETS sectors. Oversupply prevents a carbon price sufficiently high to provide real incentives for industry to invest in low-carbon solutions.

Member-state projections show that total emissions in non-ETS sectors are expected to remain lower than the 10 per cent target until 2020 (EEA, 2014b).[25] Some, like Cyprus, will probably surpass their targets; others, like the United Kingdom, are just on track, whereas Germany, Luxemburg and Poland are not on track to meet their ESD targets. Progress on RES also varies significantly. In Germany, an ambitious RES policy based on feed-in tariffs has brought a surge of new decentralized renewable electric power production.[26] Still, emissions are increasing because of imports of cheap hard coal from the United States and slow uptake of renewables in the heating and transport sectors. Conversely, the energy mix in the Netherlands is based almost entirely on fossil fuels, and new coal-fired power plants have recently been constructed. In 2013, the Dutch share of renewables in final energy consumption was only 4.5 per cent, based largely on co-firing biomass and coal, and still a long way from its EU target of 14 per cent by 2020 (Gulbrandsen and Skjærseth,

2014). Co-firing does not contribute to 'lock out' coal from the energy mix, and the RES component may prove short-lived if technology-neutral RES subsidies end before 2020 and coal becomes cheaper than biomass. Thus, significant progress has been made towards the targets – but the EU climate and energy package has hardly been a 'game changer' towards a collective EU low-emission economy. In the ETS sectors, emissions have dropped by 19 per cent compared to 2005, due mainly to the economic recession since 2008 as well as to the use of emissions reduction credits from outside the European Union.

The 2030 EU climate and energy policy framework needs to be specified by new legislation that must be implemented by the member states. This legislation will determine the probability of meeting the new 40 per cent reduction target, disregarding economic and technological changes unrelated to this policy. For example, strengthening the ETS by increasing the annual reduction factor from 1.74 per cent to 2.2 per cent is the most specific measure adopted by the European Council. Even with this reform in place, it will take a long time to 'eat up' the allowance surplus after 2021.[27] The EU has agreed on new legislation for a market stability reserve (MSR) to address the surplus problem and to strengthen resilience to economic fluctuations by automatically adjusting the supply of allowances to be auctioned. Projections indicate that the allowance price will on average be 40 per cent lower between 2021 and 2030 without the MSR, reaching about EUR 30 by 2030 with the MSR (Fjellheim, 2014). That is almost the same level as in spring 2008.

We can conclude that the EU is on track towards 2020 targets – but progress has been uneven in the member states, and the climate and energy package has not yet placed them collectively on a new 'green' pathway towards decarbonization. For 2030, a new domestic 40 per cent GHG reduction target has been adopted. From a policy perspective, achievement will depend on the ambitiousness of new legislation based on the framework and subsequent implementation by the member states.

CONCLUSIONS

EU climate policy has become gradually more ambitious in goals and policy instruments, strengthening the credibility of the EU's leader-by-example ambition in the international climate negotiations. The chapter has explored how the EU has been able to adopt an increasingly ambitious climate policy based on a modified supply–demand policy approach. Energy security concerns have provided a *need* for energy efficiency and more energy production. However, given differing interests among the member states and within the Commission, there is still a long way from energy needs to climate *action*. The institutional setting has provided an enabling context by stimulating consensus seeking and long-term policy development. A major obstacle throughout has been getting the new low-income member states to agree on a more ambitious climate policy.

Explanations for change concern how policies and issues were combined in new ways that shared the costs, promoted new low-carbon opportunities and gave something to all the major 'veto players' involved in policy-making. Policy *content* in the form of a package of climate and energy policies became a necessary condition for change. In the policy initiation phase, differing interests within the Commission and among the member states were combined by integrating climate and energy policies. The comprehensive package designed by the Commission provided ample room for exchanging of concessions among member states across various policies in the negotiations phase. Still, Poland, as the informal leader of other CEECs, was not happy with the outcome. During the implementation phase, synergies and compensation that had promoted EU-level agreement proved hard to reproduce at the national level. An ambitious EU climate policy was seen as a potential threat to energy security to the extent it threatened coal, shale gas exploration or affordable electricity prices. The new climate and energy policy framework for 2030 reflects varying experiences with implementation of the climate and energy package. This new EU 2030 framework represents a 're-packing' compromise to satisfy the main veto players, including substantial concessions given to Poland and other CEECs.

The linkage between climate and energy policies has not come from societal demands, but broad support for EU-level climate policies has been important for legitimizing the decisions taken by EU leaders. The package was initiated despite low support for more ambitious EU-level renewable and energy efficiency policies, and in some member states there is only a weak link between public perceptions of climate action urgency and government positions.

Significant progress has been made towards the EU targets, but the record in the member states has been uneven. The climate and energy package has not been a 'game changer' that could collectively put the member states on the path towards a low-emission economy. Whether the 2030 EU climate and energy policy framework will manage to achieve a low-emission economy depends on new legislation that must be implemented by the member states and on economic and technological changes unrelated to this policy. These factors are what will largely determine whether the new 40 per cent reduction target will be met.

NOTES

* The author thanks Torbjørg Jevnaker, Inga Ydersbond, Olav Schram Stokke and Steinar Andresen for their constructive comments in preparation of this chapter.

1. Because of length constraints, there is less focus on non-state actors such as industry. Non-state actors have played an important role in EU climate policy. See, for example, Skjærseth and Wettestad (2008) and Skjærseth and Eikeland (2013).
2. Within certain limits, external credits from investing in more affordable projects abroad could be used by companies to comply with the EU ETS.
3. These included a regulation covering emissions from new cars, which applied primarily to member states with car manufacturers, and a directive on fuel quality, including a required reduction of the carbon footprint of road fuels from well to wheel, which applied primarily to oil companies.
4. This is to be achieved by changing the annual ETS cap from today's 1.74 per cent to 2.2 per cent, and by a market stability reserve to address the accumulating surplus of emission allowances.
5. In addition, interconnection became a target in the EU's climate and energy policy for the first time.
6. Note that societal demands are included only as they are expressed in Eurobarometer polls. These polls can only provide an indication of pressure and resistance from 'below'. A more comprehensive study of societal demands would need to include organized interests and how organizations express their interests within the European Union.
7. Oil and hard coal are traded in liquid international markets with multiple suppliers and a well-functioning transport infrastructure, mainly maritime.
8. However, resilience has improved (storage, reverse flows and interconnectors) and winter supply disruptions pose a particular challenge for only certain countries – Bulgaria, Romania, Hungary and Greece.
9. In Poland, indigenous coal accounts for nearly 90 per cent of the country's electricity production and 50 per cent of its total CO_2 emissions.
10. The Court will not be dealt with here, as its primary functions involve actions regarding member states' non-compliance and review of legislative acts.
11. The biggest countries have the most votes. A *qualified* majority is achieved when a majority (sometimes two-thirds) of the 28 member states vote in favour, at least 260 of the possible 352 votes are cast, *and* the majority represents at least 62 per cent of the total population.
12. Existing voting rules are based on the Nice Treaty. The new Lisbon voting rules entered into force in November 2014, reducing the threshold for a qualified majority and allocating voting weights in direct proportion to member states' population. Until 2017, however, any member state can request that a Council decision be adopted according to the old Nice rules. See www.frankhaege.eu/category/blog-tags/research/eu-politics/council-eu (accessed 8 December 2014).
13. See the EU Transparency Register: http://ec.europa.eu/transparencyregister/public/homePage.do?redir=false&locale=en (accessed 27 January 2015).
14. Turnout for these elections is low (42.54 per cent in 2014), compared to the turnout for national elections.
15. By 'veto players' is meant a set of specific individuals or collective actors whose agreement is necessary to change the status quo.
16. Largely swept under the carpet were potential conflicts like downward pressures in carbon prices caused by renewables and energy efficiency measures in the ETS sectors.
17. They also felt poorly informed – 30 per cent thought that CO_2 emissions had only a marginal impact on climate change.
18. One example is the failed EU carbon/energy tax.
19. See also http://ec.europa.eu/clima/policies/ets/reform/index_en.htm (accessed 22 December 2014).

20. However, Poland has established a Green Investment Scheme funded from the sales of surplus assigned amount units (AAUs) as part of the Kyoto Protocol.
21. Opposition from energy-intensive industry has, however, been modified by the opportunity to compensate for indirect costs from the EU ETS.
22. Emissions from sectors outside the ETS would be cut by 30 per cent below 2005 levels, through effort sharing among member states. This is a significantly larger share than that called for in existing policies, whereby 10 per cent should be cut in the non-ETS sectors and 21 per cent in the ETS sectors by 2020.
23. Belgium, Denmark, Estonia, Finland, France, Germany, Italy, the Netherlands, Portugal, Slovenia, Spain, Sweden and the United Kingdom.
24. 'The Visegrad Group Countries, Romania and Bulgaria, Joint Paper on the EU Climate and Energy Framework 2020–2030', May 2014, undated, accessed 25 April 2015 at http://www.mos.gov.pl/g2/big/2014_05/1ee07c384a09ab47f346d05eac59e11b.pdf.
25. Energy efficiency measures in the residential and service sectors are expected to contribute three-quarters of the projected savings in the ESD sectors. Emissions from the transport sector are projected to increase.
26. In electricity production, the RES share increased from about 6 per cent in 2000 to 30 per cent in 2014.
27. With this change, the cap will be about 9 million tons lower than today's cap.

REFERENCES

Braun, J. (2011), 'EU energy policy under the Treaty of Lisbon rules', EPIN Working Paper No. 31, February.

Commission of the European Communities (2005), *Doing More with Less*, COM(2005) final, 22 June.

Commission of the European Communities (2006), *Green Paper on a European Strategy for Sustainable, Competitive and Secure Energy*, COM (2006) 105 final, 8 March.

Commission of the European Communities (2007a), *Limiting Global Climate Change to 2 Degrees Celsius: The Way Ahead for 2020 and Beyond*, COM (2007) 2 final, 10 January.

Commission of the European Communities (2007b), *An Energy Policy for Europe*, COM (2007) 1 final, 10 January.

Commission of the European Communities (2008), *Impact Assessment*, document accompanying the package of implementation measures for the EU's objectives on climate change and renewable energy for 2020, SEC (2008) 85/3, 23 January.

Commission of the European Communities (2011a), *A Roadmap for Moving to a Competitive Low Carbon Economy in 2050*, COM (2011) 112 final, 8 March.

Commission of the European Communities (2011b), *Energy Roadmap 2050*, COM (2011) 885 final, 15 December.

Commission of the European Communities (2012), *The State of the European Carbon Market in 2012*, COM (2012) 652 (undated).

Commission of the European Communities (2013), *Green Paper: A 2030 Framework for Climate and Energy Policies*, COM (2013) 169 final, 27 March.

Commission of the European Communities (2014a), *European Energy Security Strategy*, COM (2014) 330 final, 28 May, Brussels: European Commission.

Commission of the European Communities (2014b), *In-depth Study of European Energy Security*, SWD (2014) 330 final/3, 2 July, Brussels: European Commission.
Commission of the European Communities (2014c), *A Policy Framework for Climate and Energy in the Period from 2020 to 2030*, COM/2014/015 final, 22 January, Brussels: European Commission, accessed 3 November 2014 at http://ec.europa.eu/clima/policies/2030/documentation_en.htm.
Council of Ministers, Poland (2008), *The National Environmental Policy Plan for 2009–2012, with its 2016 Outlook*, Warsaw: Council of Ministers.
Egenhofer, C., S. Kurpas, P.M. Kaczynski and L.G. van Schaik (2011), *The Ever-changing Union: An Introduction to the History, Institutions and Decision-making Processes of the European Union*, 2nd edition, Brussels: Centre for European Policy Studies.
Eikeland, P.O. (2012), *EU Energy Policy Integration – Stakeholders, Institutions and Issue-linking*, FNI Report 13/2012, Lysaker, Norway: Fridtjof Nansen Institute.
Eikeland, P.O. (2014), *Implementing the EU 2020 Energy and Climate Package in Germany: Green Champion Struggling to Adapt*, FNI Report 9/2014, Lysaker, Norway: Fridtjof Nansen Institute.
Eurobarometer (2006), *Energy Issues. Special Eurobarometer*, Brussels: European Commission.
Eurobarometer (2008), *Europeans' Attitudes towards Climate Change*, Brussels: European Commission.
Eurobarometer (2009), *Europeans' Attitudes towards Climate Change*, Brussels: European Commission.
Eurobarometer (2011), *Climate Change*, Brussels: European Commission.
Eurobarometer (2013), *Climate Change*, Brussels: European Commission.
European Council (2007), *Presidency Conclusions from European Council 8 and 9 March 2007*, 2 May, Brussels: European Council.
European Council (2009), *Presidency Conclusions, 29/30 October*, 1 December, Brussels: European Council.
European Environment Agency (EEA) (2013), *Annual European Greenhouse Gas Inventory 1990–2011 and Inventory Report 2013*, accessed 2 February 2015 at www.eea.europa.eu/publications/european-union-greenhouse-gas-inventory-2013.
European Environment Agency (EEA) (2014a), 'EEA greenhouse gas – data viewer', accessed 2 February 2015 at www.eea.europa.eu/data-and-maps/data/data-viewers/greenhouse-gases-viewer.
European Environment Agency (EEA) (2014b), *Trends and Projections in Europe 2014: Tracking Progress towards Europe's Climate and Energy Targets for 2020*, European Environment Agency Report No. 6, Copenhagen: EEA.
European Parliament (2007), *European Parliament Resolution on Climate Change*, P6_TA (2007) 0038, 14 February, Brussels: European Parliament.
European Renewable Energy Council (EREC) (2007), *EREC Position on the Need for Sectorial Targets*, Brussels: European Renewable Energy Council.

Eurostat (2014), 'Gini coefficient of equivalised disposable income' accessed 2 February 2015 at http://ec.europa.eu/eurostat/tgm/table.do?tab=table&init=1&language=en&pcode=tessi190&plugin=1.
Fjellheim, H. (2014), 'Reform av kvotesystemet: EUs viktigste klimapolitiske veivalg' [Reform of the quota system: the main EU climate policy choices], in A. Bjartnes, O.A. Øvrebø and P. Michelson (eds), *Klimapolitikk i krysspress*, Report 3/2014, Bergen, Norway: Norsk Klimastiftelse, pp. 31–4.
Green Growth Group (2014), 'Green Growth Group Ministers' statement on climate and energy framework for 2030, joint statement on 3 March 2014', accessed 2 February 2015 at https://www.gov.uk/government/news/green-growth-group-ministers-statement-on-climate-and-energy-framework-for-2030.
Gulbrandsen, L.H. and J.B. Skjærseth (2014), *Implementing the EU 2020 Climate and Energy Package in the Netherlands: Mixed Instruments, Mixed Results*, FNI Report 10/2014, Lysaker, Norway: Fridtjof Nansen Institute.
Hix, S. (2005), *The Political System of the European Union*, New York: Palgrave.
Ibec (2013), *Summary of Member State Responses to the 2030 Green Paper for a Climate and Energy Framework*, October, Dublin, Ireland: Ibec.
Jevnaker, T. (2014), *Norway's Implementation of the EU Climate and Energy Package: Europeanization or Cherry-picking?*, FNI-Report 7/2014, Lysaker, Norway: Fridtjof Nansen Institute.
Oberthür, S. and M. Pallermarts (eds) (2010), *The New Climate Policies of the European Union*, Brussels: VUB Press.
Skjærseth, J.B. (1994), 'The climate policy of the EC: too hot to handle', *Journal of Common Market Studies*, **32**(1), 25–45.
Skjærseth, J.B. (2013), *Unpacking the EU Climate and Energy Package: Causes, Content, Consequences*, FNI Report 2/2013, Lysaker, Norway: Fridtjof Nansen Institute.
Skjærseth, J.B. (2014a), *Implementing EU Climate and Energy Policies in Poland: From Europeanization to Polonization?* FNI Report 8/2014, Lysaker, Norway: Fridtjof Nansen Institute.
Skjærseth, J.B. (2014b), 'Linking EU climate and energy policies: policy-making, implementation and reform', *International Environmental Agreements: Politics, Law and Economics*, published online 2 October 2014, DOI: 10.1007/s10784-014-9262-5.
Skjærseth, J.B. and P.O. Eikeland (2013), *Corporate Responses to EU Emissions*, Aldershot, UK: Ashgate
Skjærseth, J.B. and J. Wettestad (2008), *EU Emissions Trading: Initiation, Decision-making and Implementation*, Aldershot, UK: Ashgate.
Skjærseth, J.B. and J. Wettestad (2010), 'Fixing the EU emissions trading system? Understanding the post-2012 changes', *Global Environmental Politics*, **10**(4), 101–23.
Skjærseth, J.B., G. Bang and M. Schreuers (2013), 'Explaining growing climate policy differences in the European Union and the United States', *Global Environmental Politics*, **13**(4), 61–79.
Skjærseth, J.B., P.O. Eikeland, L. Gulbrandsen and T. Jevnaker (forthcoming), *Linking EU Climate and Energy Policies: Policy-making, Implementation and Reform*, Cheltenham, UK and Northampton, MA, USA: Edward Elgar Publishing.

Tsebelis, G. (2002), *Veto Players: How Political Institutions Work*, Princeton, NJ: Princeton University Press.

World Bank (2014), 'World DataBank', accessed 24 April 2015 at http://databank.worldbank.org/data/home.aspx.

Ydersbond, I.M. (2014), 'Multilevel "venue shopping": the case of EU's Renewables Directive', *Interest Groups & Advocacy*, **3**(1), 30–58.

5. Climate politics, emission scenarios and negotiation stances in India

Sunil Tankha and Trude Rauken

BOX 5.1 KEY FACTS ABOUT INDIA

Population 2013	1.252 billion
Population growth rate 2013	1.2%
Population density 2013	421.1 per km^2
GDP per capita 2013	1498.90 USD
GDP growth rate 2013	4.7
GINI index 2011	33.6
Access to electricity 2010	75% of the population
Land area	3 287 590 km^2
Political system	Federal parliamentary democratic republic
Subnational entities	29 states and seven union territories

Source: World Bank data (2014d, 2015a–g).

INTRODUCTION

India (see Box 5.1 and Table 5.1) is the world's third-largest emitter of carbon dioxide, and although its per capita emissions are only about a tenth of those of the United States, India's future emissions will significantly affect global efforts to combat climate change. Even under the best-case scenario, it is projected that India will emit about 5.5 billion tons of CO_2 annually by 2031, which is almost the same as what the United States currently emits annually (Planning Commission, 2006). Moreover, India is one of the countries that are likely to be most affected by climate change, with hundreds of millions of poor being very vulnerable to its impacts. Public action on climate change is therefore urgently needed, but will this be forthcoming? In this chapter, we analyse the tripod of governance capacity, developmental and energy needs, and political exigencies to understand India's climate politics and policies, and emission scenarios, and through this explain what action we can expect to come from India.

This analysis is based on the cumulative data collected by the authors over multiple and extensive field studies conducted over a period of several years. During these field studies, which focused on climate change (Tankha and Rauken, 2013) and reform in the energy and water sectors in India (Tankha et al., 2010; Tankha and Fuller, 2010), the authors conducted semi-structured interviews with dozens of key informants and stakeholders at all levels of government and with climate-vulnerable communities. The data from these interviews are supplemented by extensive reviews of official documents, news reports, parliamentary proceedings and the academic literature. We also review the country's National Action Plan for Climate Change (NAPCC) to understand official attitudes and where and why India is placing its priorities with respect to the challenges created by climate change. In addition to analysing the standard political economy variables, we cast a broader net to look at several other factors affecting environmental politics and practices in India, and in doing so to develop a more detailed narrative of competing values, aspirations and capabilities.

The chapter is divided into five sections. Following the introduction we describe India's preferred approaches in the United Nations Framework Convention on Climate Change (UNFCCC) negotiations and the domestic actions contained in the NAPCC. To explain these we need to look at developmental needs and governance structures. Hence, in the third section we examine India's energy demands and supplies to understand what constrains and conditions India's political and economic responses to climate change mitigation. In this section we also deal with the traditional political economy variables in some detail. This includes a brief analysis of India's political and administrative decision-making arrangements, political priorities and styles, and administrative cultures and incentives. We conclude the section by analysing the main drivers of current energy policy and environmental governance in terms of both the supply of environmental policies by government and the demand for these policies from society. In the fourth section, we chart India's probable policy trajectory, based on our analysis. Finally, the fifth section concludes with some observations about the kinds of strategies to address climate change that are likely to find traction in India.

Table 5.1 GHG emissions and energy mix in India

Total GHG emissions by year	1990	2005	2012
	1354.5 $MtCO_2e$[a]	2104.7 $MtCO_2e$[b]	3013.8 $MtCO_2e$[c]
GHG emissions per capita 2012[d]	2.43 tCO_2e		
GHG emissions by sector 2012[e]	Energy 69.6%, transport 7%, industrial processes 6.5%, agriculture 14.1%, waste 2.4%, bunker fuels 0.5%		
Energy mix 2012[f]	Coal 44%, petroleum and other liquids 22%, biomass and waste 22%, natural gas 7%, hydroelectric 3%, nuclear 1%, other renewables 1%		

Notes:

a. International Energy Agency (IEA) (2014).
b. IEA (2014).
c. WRI (2014).
d. Emission Database for Global Atomospheric Research (EDGAR) (2015).
e. WRI (2014).
f. US Energy Information Administration (EIA) (2014).

INDIA'S APPROACH TO CLIMATE CHANGE MITIGATION AND ADAPTATION: INTERNATIONAL AND DOMESTIC STANCES

Climate change poses two distinct challenges for developing countries. The first challenge is to understand what impacts climate change will have on the country. The second challenge is to find the resources to deal with the longer-term impacts of climate change when immediate needs are so critical and pressing.

When a developing country is large and growing, as is India, a third challenge arises: how to protect economic growth against climate-related political limits. By this we refer to the situation wherein a country's total greenhouse gas (GHG) emissions reach such a level that other countries are no longer willing to accept non-participation in mitigation efforts. Being the world's third-largest emitter behind China and the United States, India is now facing increasing international pressure to commit to reduce GHG emissions (Gallucci, 2015).

India's stance towards mitigation is that since climate change is mainly a consequence of the wealthier countries' historic and current emissions, it is they who should be responsible for mitigating it and for helping to pay for adaptation responses in developing countries. This stance, which overall is shared across the political spectrum in India, is based on the principle of common but differentiated responsibility (CBDR) and respective capabilities. Its underlying philosophy is that while all states are responsible for contributing to climate change, some states bear a disproportionate current and historic responsibility. This argument obviously places India in a favourable position as, according to one estimate, India's contribution to cumulative emissions is around 3 per cent as compared to almost 25 per cent for the United States and 13 per cent for China (Columbia University, 2014).

The initial reaction of the industrialized countries to India's CBDR stance was predictably critical. They warned that India risked being viewed as stubborn and as not taking the potential impacts of climate change on itself seriously (see, for example, *The Economist*, 2008). In response, Indian delegates began to coordinate more with other developing countries, such as the G77 and China and the Like-Minded Developing Countries on Climate Change (LMDC) group, to avoid negotiating from an isolated position and to advocate for mitigation targets based on environmental equity (Kallbekken et al., 2014). As a result, India has now emerged as a key player and a leading voice for the developing/G77 countries in the North–South divide in emission control negotiations (Sengupta, 2012).

Indian negotiators are inclined to change their minds on operational procedures when they are complementary to India's basic positions and interests. For example, India reversed its initial opposition to the introduction of the market mechanisms in the Kyoto Protocol and later became enthusiastic about them, for example, embracing the Clean Development Mechanism (CDM) because it began to see an economic opportunity in it. Indeed, India is now host to a large proportion of ongoing CDM projects and has attempted to use them to support domestic energy and industrial priorities, including hydro and nuclear power (Phillips and Newell, 2013). This kind of strategic bargaining – undertaking climate change mitigation initiatives but explicitly in exchange for financial and technological resources – has allowed for protecting domestic economic interests while responding to pressure from the international community (Dubash, 2013).

However, when the negotiators' flexibility in mitigation discussions has been perceived to conflict more fundamentally with India's economic growth interests, the domestic political constituencies have mobilized strongly against them. For example, at the 16th Conference of Parties in Cancun, India's Minister of Environment and Forests, Jairam Ramesh, indicated that India could agree to binding emission targets in the future. This immediately led to strong criticism of him in India's Parliament, with the opposition accusing the government of succumbing to foreign pressure and compromising national sovereignty (NDTV, 2010). Ramesh was then forced to issue a clarification in Parliament, essentially back-tracking from his position (Parliament of India, 2010). He was shortly thereafter promoted out of the position and made Minister for Agriculture and Rural Development (Gupta, 2011; GoI, 2011a).

More representative of India's position is the then Prime Minister Manmohan Singh's statement to Parliament after the G8 meetings in Italy in 2013. He assured Parliamentarians that he had informed the G8 and G5 countries that India recognizes 'an obligation to preserve and protect our environment, but climate change cannot be addressed by perpetuating the poverty of the developing countries' (Parliament of India, 2009). He also said that India was 'willing to do more provided there are credible arrangements to provide both additional financial support as well as technological transfers from developed to developing countries', indicating that India will not compromise on economic growth, but would take mitigation actions as long as they are facilitated by finances and technology from industrialized countries. India sees these kinds of mitigation actions as contributing to growth and competitiveness rather than detracting from them.

India started paying more attention to climate change and in particular to its vulnerability to climate change when the Ministry of Science & Technology developed the NAPCC just before the launch of the Intergovernmental Panel on Climate Change Fourth Assessment Report in 2007. As they sat down to write the NAPCC, Indian formulators noted that regardless of who was responsible, India's main challenge was going to be adaptation as the available research predicted increased variability in the monsoons, serious changes in the Himalayan ecosystems, reduced agricultural productivity and health impacts (MoEF, 2004, 2012).

As the lists of potential impacts were developed, it became apparent that while climate change would not create fundamentally new challenges for governance and administration, it would greatly exacerbate existing challenges because the unpredictability and frequency of extreme events would increase. Thus, India did not need to develop entirely new governance capabilities as much as it had to strengthen existing capacities and develop new and more effective ways to address existing challenges (Tankha, 2009).

Adaptation concerns received greatest emphasis in the NAPCC. The document organizes India's main climate-related challenges under eight mission areas of which the three largest ones – the National Water Mission, the National Mission for Sustainable Agriculture and the National Mission for Sustaining the Himalayan Ecosystem – are focused on adaptation measures. Two of the missions – National Mission for Enhanced Energy Efficiency and the National Mission on Sustainable Habitat – are focused on increasing energy efficiency while another two smaller missions – National Solar Mission and National Mission for a Green India – have a mitigation scope. The final mission – the National Mission on Strategic Knowledge for Climate Change – supports both adaptation and mitigation goals.

Summarizing, India's external climate politics aim at shielding itself from binding growth-suppressing emission-reduction commitments while its domestic climate change focus is on enhancing climate change adaptation capabilities. However, while the government has energetically focused on external politics, internally its actions on adaptation lack urgency. In addition to a general limitation of resources, we found that there was a frustrating lack of knowledge and understanding about climate change impacts at local levels, where adaptation action is essential (Tankha and Rauken, 2013).

EXPLAINING POLICY TRAJECTORIES

Although India's development goals trump climate change concerns, India's size and potential future emissions dictate that progress on climate change mitigation must somehow include India. India's willingness to contribute will largely be determined by its energy needs and options. Thus, in this section we first explain and discuss the *fundamental* parameters in terms of natural resource endowment and institutional arrangements around energy and then the *driving* forces in terms of politics and administration.

Fundamental Parameters

India is the world's fourth-largest energy consumer after the United States, China, and Russia, but has a unique energy profile. The most telling statistic is that almost a quarter of the energy consumed in India comes from biomass and waste. About 70 per cent of Indian households live in rural areas, and almost two-thirds use firewood for cooking while most of the others use crop residue and dung. Even among urban residents, a fifth use firewood for cooking and about 3 per cent still use crop residue and dung (CensusInfo India, 2011). This fuel poverty is the most striking feature of India's current energy use, and a move away from this situation will be the major factor marking India's energy transition.

Material parameters: natural resource endowments

Coal is the biggest source of energy in India and in 2013 it provided more than half of the total energy consumed. Crude oil accounted for a little less than a third of energy requirements, followed by natural gas (7 per cent), hydroelectric (5 per cent) and nuclear and other renewables (around 1 per cent each) (EIA, 2014). India has the world's fifth-largest coal reserves and is the third-largest producer and consumer thereof. Despite its coal dependency, coal mining is badly managed, riddled with corruption and unable to meet the increasing demand for power generation. Consequently, India has begun to import coal in increasing quantities and currently imports about a third of its coal consumption (ibid.). Despite these problems, diversification away from coal is difficult because none of the alternative fuels present an attractive or feasible policy choice.

Oil is a necessary evil for policy-makers because almost all the oil India consumes must be imported and the cost amounts to over 30 per cent of India's total imports. The import needs are growing and security of supply and vulnerability to foreign exchange rates remain important concerns.

Natural gas is not viewed as a desirable alternative on a large scale as it too must be imported and an increase in natural gas consumption would expose India to foreign exchange outflow risks. Supply concerns are also relevant. Although natural gas can be sourced massively from neighbouring countries, the region's political instability means that the pipeline infrastructure required is infeasible.

Hydroelectric power is a desirable emissions-reducing technology because it is economically competitive. However, large hydroelectric power projects require the displacement of hundreds of thousands of people. Issues related to land acquisition, and the compensation for and resettlement and rehabilitation of people involuntarily displaced by dams and other public infrastructure projects have been extremely controversial in India and thus politically difficult.

Nuclear power is another alternative, and installed capacity is projected to increase following the 2005 agreement on civilian nuclear cooperation signed between India and the United States, which brought India into the Nuclear Suppliers Group and gave it access to technologies, components and uranium. While the government projects that nuclear power will supply 25 per cent of the electricity requirement in 2050, the long time horizon of this planning and India's frequent slippage on other targets indicate that this is doubtful.

Wind and solar power are the green technologies, along with small run-of-river hydropower, which are the most desirable because they have the fewest negative externalities. While small hydropower plants are environmentally preferable as well as economically viable, their potential is limited by the availability of suitable sites. The other two technologies are currently economically uncompetitive and rely on subsidies and incentives that are beginning to create governance problems in terms of incentives management (Chaudhary et al., 2014; Altenburg and Engelmeier, 2013). Nevertheless, these technologies are spreading. India already has the world's fifth-largest installed capacity of wind power, totalling over 20 GW in 2013. Solar installation has also increased rapidly in India since 2011 and at the end of 2013 accounted for approximately 1 GW. The GoI's National Solar Mission targets a total installed capacity of 22 GW by 2020 (GoI, 2011c). However, since both wind and solar power are still far from being economically competitive, current installed capacity is mostly a result of government subsidies and incentives, for example, guaranteed tariff schemes ensuring an attractive return on investment to the producers.

Technological advances and economies of scale in production will help prices of wind and solar power fall even further and will allow greater penetration into the supply chain, but until these technologies become competitive with coal, and given India's natural endowments and economic concerns, there are few incentives to move away from coal. Deviating from this trend will require considerable political commitment (to incur the cost) and administrative expertise (to manage complex market subsidies and improve the functioning of India's mostly state-run energy sector). We now discuss these factors.

Institutional parameters: the political system

India is a large and diverse country, with over a billion people and over 20 official languages. Politically, it is organized as a federal republic but with a strong centralizing tendency. Currently, India is subdivided into 29 states and seven union territories, the latter being small regions governed directly by the central government (see Box 5.1 above).[1]

The Indian Constitution divides legislative and administrative responsibilities between the centre and the states into three lists: the Union List – the exclusive prerogatives of the central government (e.g., defence and telecommunications); the State List – the prerogatives of the state governments (e.g., agriculture and public health); and the Concurrent List – items where both the central and state governments may legislate and administer, though in cases of conflict, laws passed by the central government prevail. The Concurrent List contains environmental issues such as forests and wildlife protection as well as electric power and water.

There are very few veto points in the Indian political system. As it is a parliamentary democracy where the leader of the largest party or coalition of parties in the lower house of Parliament is selected as prime minister and in turn appoints the Cabinet of Ministers, there is little separation between executive and legislative powers in India. Moreover, the ruling party or coalition can pass almost all legislation at will since by definition it has an absolute majority in Parliament and since a constitutional amendment in 1985 allows the disqualification of members of Parliament who abstain or vote against their party's directions.[2] Hence, the only practical limit on a government's legislative powers is the need to keep coalition members, if necessary and applicable, on board. This is sometimes an issue as India's politics are a vibrant multi-party affair.

Nevertheless, while the Constitution appears to delineate responsibilities quite clearly, in practice the central government, mainly through its control over public finances, can significantly influence state government actions even for subjects on the State List. The Planning Commission of India, a central government agency, used to determine the allocation and control the disbursement of funds to the states, and although the agency was recently scrapped in 2014,[3] it is expected that some other agency will continue this centralized practice. Thus, even though states are guaranteed funding from the centre, it often comes with detailed instructions. At the state level, this remains a contentious issue, especially in states headed by parties who are not in power at the centre.

One example is the preparation of the State Action Plans for Climate Change (SAPCC). These are direct responses to national directives to prepare these plans, based on local needs and priorities. These plans must be approved by the central government; so although the process has the veneer of decentralization, it could be more accurately characterized as delegation or deconcentration. The lack of veto points and the centralization in the Indian political system lead to the fact that central directives often pass without adequate local-level support and thus either are not seriously undertaken or are actively or passively sabotaged. Thus, the policy analyst should not assume that adopted policies translate neatly into implementation.

Like the material parameters above, the political parameters strongly influence the energy sector. One of the biggest political issues, which has wide ramifications in all areas affected by climate change, including agriculture, is the question of energy subsidies. Total energy subsidies are currently at US$17 billion (IISD, 2012). India subsidizes liquefied petroleum gas (LPG) and kerosene because they are used for cooking, and diesel because it is used in transport, including goods transport. Energy subsidies are also one of the biggest drains on the state electricity companies. These companies provide free or almost free power to agricultural consumers who consume about a third of the electricity supply but provide only 4 to 5 per cent of power sector revenues.

The politics of energy subsidies are intense in India. There is huge resistance to price increases for power supplied to farmers not only from the agricultural lobby, but indirectly also from urban India, which is hugely concerned about increases in the price of basic staples. Fuel subsidies are also politically important because about a third of India's population falls below the nationally defined poverty line, which was just recently redefined as living on less than a dollar a day – 32 rupees a day in rural areas, and 47 rupees in urban areas (Singh, 2014). Every time the government proposes a reduction in fuel subsidies, opposition politicians (and sometimes even coalition partners) agitate for a rollback of the price increases. A partial rollback is, in fact, often done. Thus, using price signals as a means of reducing consumption of fossil fuels or of encouraging the transition to more fuel-efficient alternatives is not feasible. Efforts by the central government to reduce fuel subsidies in 2014 were more successful, but these were facilitated by falling international prices for fuel.

Arguments that fuel subsidies do not reach the very poor (because they are still too poor to afford public transport or gas or kerosene cooking stoves) are politically irrelevant. Many in the middle class live financially precarious lives in which increases in the prices of basic necessities do represent a hardship significant enough that they would or could be politically mobilized. Better targeting of subsidies is theoretically desirable, but given the current informational infrastructure, it would be administratively burdensome and would likely be thwarted.

Institutional parameters: the administrative system

India's political centralization is reflected also in the administrative set-up. The apex of the central service is the Indian Administrative Service (IAS) supplemented by various other central services such as the Indian Foreign Service, Indian Postal Service, Indian Forest Service and Indian Police Service. The states have their own civil services established on a similar basis, but are considered less prestigious than the IAS because the more important posts in the state administration are reserved for the latter.[4]

The Indian civil services follow a classic Weberian model, from which they derive their inspiration and strength but also their fundamental weakness. At the lower levels, all of the ministries, departments and state-owned enterprises have technically capable permanent staff that provide a fair degree of continuity in their organizations' operations, but the top level of the administrative set-up is composed of bureaucrats who are skilled generalists but possess little sectorial expertise. Given their frequent rotation (their average tenure in a post is less than three years), they have little opportunity to develop profound sectorial expertise. While their initial selection is meritocratic and job tenure is secure, civil servants have limited functional autonomy with respect to the elected political leaders. The latter have the power to appoint and transfer bureaucrats from their posts at will, and exercise this power frequently. Politicians, who are known to micromanage all government affairs, insist that bureaucrats keep them constantly updated on operational issues, and routinely monitor and interfere in bureaucratic decisions and processes. Moreover, senior bureaucrats delegate little because politicians hold them responsible for any problems. Political loyalty might be expected to appear given the close control exercised by politicians over bureaucrats, but in practice, while favourites may sometimes emerge, the administrative services are fundamentally apolitical and execute whichever policies the elected representatives decide upon.

The combination of political micromanagement, the pressure for administrative responsiveness to the political leaders and civil service tenure has led to concentration of workloads at the upper levels. This administrative congestion means that little time is available for strategic thinking and systemic reform (Tankha and Rauken, 2013). Consequently, rather than attempt to fundamentally reshape public governance structures, reformers are forced to outsource functions and rely on technological bypasses to administrative pathologies. This, of course, limits the scope of basic reform and addressing emerging policy challenges related to climate change.

As in other arenas of governance in India, energy governance is characterized by blurred and overlapping boundaries and fragmented authority and responsibility. The Ministry of Petroleum & Natural Gas is responsible for oil and natural gas exploration and production and for the distribution, marketing, import, export and conservation of oil and natural gas (GoI, 2014c). The Coal Ministry holds the overall responsibility for development and exploration of coal and lignite reserves, as well as for pricing and distributing coal (GoI, 2014a). The Ministry of Power (MoP) is responsible for policy, planning and project implementation in the electricity sector and for coordination of issues relating to energy policy, including rural electrification (GoI, 2014d). Finally, the Ministry of New and Renewable Energy, established in 2006, is a nodal ministry responsible for all matters that relate to green energy sources (GoI, 2014b). There are large areas of overlap among the four ministries that oversee the energy sector, and this overlap leads to a situation whereby coordination is lacking and strategic planning suffers. Furthermore, under the ministries are a number of state-owned enterprises that operationally execute the government policies, for example, the Oil and Natural Gas Commission and the Indian Oil Corporation. In the power sector, electricity generation is split between the nationally owned power generation companies and the State Electricity Boards (SEBs), while power distribution is exclusively the latter's responsibility. Some distribution areas have recently been privatized.

Because of political interference and administrative ineffectiveness, the energy sector, especially the power sector, remains mired in inefficiency, with about one-fourth of the electricity produced lost in transmission and distribution and about 10 per cent unpaid for. This inefficiency means that less electricity is produced in India than there would be if the sector were financially solvent. In addition, a significant amount of the power produced is wasted, leading to unnecessary emissions. Moreover, the coping strategies of consumers increase emissions as they entail installing diesel-fuelled back-up generation units, which are highly inefficient in terms of emissions. As a result, electricity production trails demand and the energy intensity of the GDP is higher than it could be.

Institutional parameters: the judicial system

Another important factor that has emerged in recent years in the arena of environmental policy is the court system. While there is little separation between the legislative and executive powers in India, the judiciary is quite independent and uses broad powers of judicial overview to call the government to account even on policy and administrative matters. Through a provision in the law that allows for litigation in the public interest, environmental groups and engaged citizens have petitioned the Supreme Court and High Courts of India to force governments to enforce applicable laws and pass new administrative measures that improve environmental quality. The landmark case in this arena was one brought in 1985 by M.C. Mehta against the Government of Delhi. Mehta sued the Government of Delhi, arguing that air quality in New Delhi was a hazard to human health, largely caused by dirty diesel used in public transport, and that this, in effect, violated the constitutional guarantee to life. The Supreme Court agreed, ruling that all public transport in the National Capital Region needed to run on cleaner-burning natural gas. In effect, this ruling required the retrofitting of all buses, taxis and auto-rickshaws. Although the Government of New Delhi resisted, the Court compelled the authorities to issue the relevant orders to retrofit the transport fleet and construct the necessary infrastructure (Dholakia et al., 2013). Eventually, fuel conversion for public transport was extended to several other metropolitan areas. The success of this petition shows that in many cases what the government claimed was economically or administratively impossible was in fact not so, and it also shows that a rights-based approach to governance or overcoming governance deficiencies is possible in India.

Generic Forces

The central government's response to climate change mitigation is conditioned by India's energy supply sources and energy governance. In the previous section, we noted three salient points: (1) that the current supply sources are dominated by fossil fuels; (2) that competitive politics means that delivering on economic growth and populist pricing in public services are likely to trump other concerns; and (3) that energy governance is weak, leading to inefficiencies. In this section, we discuss the supply of government policies targeted at climate change as well as the social demands for climate-related policies.

Governmental Supply

The GoI's two main thrusts towards mitigation and emission reductions are to increase investments in renewable energy, specifically wind and solar power, and to increase energy efficiency. Wind and solar power are targeted to reach installed capacities of 40 GW and 20 GW, respectively, by 2020 (GoI, 2011c). The main government policies to achieve these targets focus on private sector investments through tax breaks, accelerated depreciation allowances and preferential procurement. At the national level, proposals are being discussed to require SEBs to purchase a minimum amount of their grid requirements via renewable energy sources, and to facilitate this a trading scheme of renewable energy certificates is under development. There is also a proposal pending in Parliament to levy a tax on coal to fund development of cleaner energy alternatives.

The latter initiative, of energy efficiency, is under the MoP and it has established a target to achieve 10 GW of energy efficiency savings by 2020 (GoI, 2008). Both of these approaches can be expected to deliver only modest results as even under the best-case scenarios, emissions by India will increase substantially. In 2006, the Planning Commission released India's Integrated Energy Policy. This document presents 11 scenarios of fuel mix, ranging from the most economical, in which coal dominates, to the most progressive, in which not only the full potential of hydropower is realized (150 GW), but also nuclear generation capacity increases significantly (to 63 GW). The policy provides for greater efficiency in all areas including coal-fired power plants, and provides for increased railway transport and increased fuel efficiency standards for all motor vehicles. The difference between scenario 1 and scenario 11 in terms of CO_2 emissions is approximately 35 per cent: 5500 million tons per year in 2031 under scenario 1 compared to 3900 million tons under scenario 11 (Planning Commission, 2006). The only possibility for a radical change in these trends is a complete transformation in the economics of renewable energy production through technological breakthroughs or economies of scale. The government supply of relevant policies and initiatives is therefore limited.

Societal Demand

It is difficult to generalize based on the limited data that is available on people's perceptions of climate change, but the consensus is that the urban educated classes have been broadly exposed to the issue. Younger and more educated people are more aware of climate change and its causes, though politically they are not yet a force for change. In addition, there is no noticeable climate change–denier sentiment.

While climate change mitigation does not directly resonate with the Indian population as a political or administrative issue, local environmental problems do, particularly increased vehicular pollution in the major metropolitan cities. New Delhi and Beijing are the world's two most polluted cities, often alternating in their ranking. All the other metropolitan cities of India are also heavily polluted. Under public pressure to do more to tackle air pollution, the Central Pollution Control Board launched a National Air Quality Index for major cities in October 2014 as a first step in developing awareness and urgency for action. Thus, although the mass media and citizens are concerned about the environment, this concern is more about local pollution problems than about global environmental problems like climate change. Although cleaner-burning energy and better transport systems may contribute to increasing the efficiency of energy use, the connection between these concerns and climate change is not necessarily made.

The situation is different in rural India. Because the monsoons provide respite from the searing heat of the Indian summer, their arrival is an eagerly anticipated annual ritual throughout India. In recent years, the arrival and departure of the monsoons have become more erratic and there is thus a widespread awareness that the weather and climate are changing. Farmers in particular are keenly attuned to the arrival and departure of the monsoons because of their impacts on crops and harvests. Because of the regularity and importance of the monsoons in Indian culture and agriculture, changes in these patterns have immediate political ramifications. In particular, the regular spate of farmer suicides, usually blamed on rainfall-related crop failures and exacerbated by their taking on debt to purchase seeds, is a recurring theme of emotional debate in Parliament (see, for example, Parliament of India, 2013). Thus, it is in rural India that the biggest concern about climate change is being expressed, and the concern is more about adaptation than mitigation.

LIKELY POLICY TRAJECTORY

What then are the driving forces behind climate change mitigation and adaptation action in India? Based on our analysis, we can distil several trends that have to varying degrees shaped and will influence India's climate change policies and negotiation positions.

First, material parameters as well as governance issues hobble India's energy sector and point to a near-term future of increased emissions (see, for example, the Planning Commission scenarios). Energy consumption will increase at accelerated rates in the coming decades. This increasing demand comes on top of the current power deficit, which averaged around 9 per cent during peak hours at the end of the 11th Five Year Plan (GoI, 2013b). Moreover, almost one-fourth of Indians do not have access to electricity, and as government programmes begin to connect these citizens, electricity consumption will increase, although at a decreasing rate (GoI, 2013a). The extent of likely consumption growth can also be estimated by noting that per capita electricity consumption in India is currently less than 1000 kWh as compared to the global average of 2600 kWh and the EU average of 6700 kWh (World Bank, 2015a). This new demand will largely be met by fossil fuels. Even though wind and solar energy are promising, they are still several years away from being deployed as the main baseload-producing technologies.

Second, domestic politics are intensely competitive and do not permit any scope for combating emissions if doing so would limit economic growth. As long as India is dependent on fossil fuels for energy production – and we currently see no indications of a massive shift away from coal – it will continue to increase its emissions at rates that reflect the maximum economic growth achievable. Facing an electorate almost half of which is below the poverty line, political parties cannot avoid the development imperative. This red line is well drawn and recognized, and we can expect substantial continuity in negotiation stances because the administrative services, which form the backbone of the Indian negotiations team, receive clear instructions from the political leadership on this issue.

Although politically a constituency for reducing emissions does not exist currently, Indian political parties are not ignoring climate change. In connection with 2014 national elections, we saw for the first time the two major political parties, BJP and Congress, paying attention to climate change issues (da Costa, 2013). Although the focus was more on impacts and vulnerability, the BJP acknowledged that building a low-carbon economy would not only mitigate the climate threat, but would also be the largest economic opportunity of the twenty-first century (ibid.). However, observers use terms like 'mixed signals' (Sharma, 2014) and 'mysterious' (Yeo, 2014) when describing the BJP government's approach to climate change, indicating that there are few concrete signs of any shift in Indian climate policy towards more ambitious emission reduction targets. Moreover, this increased attention to climate change is occurring mostly at the upper echelons of government. At lower levels, our interviews revealed that there is scant knowledge of or engagement with climate change.

But the development imperative also means that if engaging with climate change yields economic opportunity, India will make positive moves towards mitigation. For example, several Indian states, including Gujarat and Tamil Nadu, have energetically pursued policies to attract wind and solar investments. Thus, we note that one of India's strategies in climate negotiations is to push for technology transfer, which would reduce the cost of energy production as well as create business opportunities for Indian industries, much as reduced patent protection created in its pharmaceutical industries. On this front, India has had limited success. One senior administrator informed us that on his trips to discuss climate-related practices and technologies in Europe he was rebuffed every time he broached the issue of technology transfer.

Third, international pressure in climate negotiation forums will be largely ineffective in curbing emissions in India. Acceding to emission curbs in response to such international pressure has invoked accusations of failing to defend national sovereignty. The ability to exert pressure on India in international forums is in any case limited. India's market is too attractive and it is politically powerful enough to successfully resist outside pressures, either alone or in concert with other countries. For example, the European Union's recent attempt via the Emissions Trading System for Aviation to impose a carbon tax on airlines flying to Europe failed after India led 26 other nations to collectively oppose the system and threatened countermeasures, including banning EU airlines from flying over Indian airspace.

Fourth, price-based mechanisms to increase energy consumption efficiency are not likely to yield much benefit in terms of emission reductions. Paradoxically, while refusing to accept emission curbs, the political and administrative structures and their shortcomings ultimately limit both economic growth and emission increases. While subsidies may encourage some wasteful consumption, their contribution to the financial limitations of the electricity sector companies has become an effective curb on emissions as they have been unable to keep up with increasing demand and power cuts are frequent and extensive. Moreover, since a part of these subsidies is derived from setting higher prices for industrial and commercial consumers, they even create price-based incentives for the large energy consumers to become more efficient in their own consumption. Had governance weaknesses been less intense in the power sector, even more energy would have been produced and consumed in India.

Fifth, India's policy-makers are much more interested in climate change adaptation than in mitigation. The available science and recent experience indicate that in India climate change will have a large and negative impact on the poor. The two major impacts according to the available studies are first a change in the monsoons, potentially causing up to a 40 per cent reduction in agricultural output especially on non-irrigated land, and second, an increase in the frequency and intensity of extreme weather events such as cyclones (World Bank, 2013). Dealing with agricultural failures and natural disasters will be a substantial adaptation task, and indeed the task already preoccupies the political and administrative structures (Tankha and Rauken, 2013). Because erratic monsoons and natural disasters disproportionately affect the poor, a pro-poor approach to climate change is also seen as better served by more investment in adaptation than in mitigation. Thus, Indian policy-makers see relatively little need to invest scarce resources in mitigation; nevertheless, they are still not investing massively in adaptation.

Sixth, both the political and administrative authorities are interested in technically sophisticated but administratively easy solutions. The political and administrative structures of India do not lend themselves to decisive policy-making or, more importantly, to decisive policy implementation. One roadblock is the competitive political system that obstructs bold decision-making. Another is that responsibilities are fragmented and divided among various levels of government, and while the central government has concentrated decision-making powers, for implementation it must rely on other levels of government, thus blunting its effectiveness. For climate change policies, political and administrative fragmentation means that even if commitments were to be made by one level or organ of government, such commitments would not ensure compliance in other levels or organs. Because they recognize these limitations, Indian authorities are unlikely to accede to any monitoring mechanisms that are administratively intensive. For a similar reason, although the energy sector is dominated by state-owned companies, more than the public sector it is the private sector, through its policy-incentivized investments in clean energy, that is likely to be the most important and active player in clean energy development.

Finally, civil society and the general citizenry in India are not in favour of emission curbs for mitigation purposes. While a reading of the available literature may give the appearance that there is more domestic concern and appreciation for a mitigations-based approach to the climate change issue, in reality this resonates far less with mainstream Indian views. The most relevant civil society perspectives, such as the widely cited report from the Centre for Science and Environment (Agarwal and Narain, 1991), which clearly influenced the India government's position and local public opinion, focus on environmental equity and not emission reductions.

CONCLUSIONS

Climate change mitigation as well as adaptation requires large investments. Given that even under the worst-case scenarios India's per capita emissions would still be a small fraction of the OECD average, Indian policy-makers feel little historical or contemporary responsibility for global warming. For them, economic growth trumps other considerations. Although statistically India is less unequal than, for example, Brazil (its GINI coefficient is 33.9 as compared to Brazil's 54.7) (World Bank, 2014d), the contrasts are much starker because absolute poverty is much more prevalent in India. About 24 per cent of India's population lives below the international poverty line of US$1.25 per day in PPP terms as compared to less than 4 per cent in the case of Brazil (ibid.). This means that India has less capacity to reduce inequalities and less space to compromise on development priorities than the other major developing countries. Therefore, in the short to medium terms India is likely to increase its emissions through increased energy consumption, focus its efforts mainly on adaptation to climate change and stick to its CBDR stance in the UNFCCC negotiations on mitigation.

If India does increase its focus on adaptation, we might see a stronger presence in the adaptation negotiations under the Convention. Several authors (Rastogi, 2011; Atteridge et al., 2012) have highlighted the fact that India wishes to be a more consequential player in climate negotiations. Even though the former Minister of Environment and Forests Jairam Ramesh was eventually unable to radically change India's approach, India has taken on more of a deal-maker role in recent rounds of negotiations (Michaelowa and Michaelowa, 2012). While our research does not indicate that these tendencies represent a change in the fundamental dynamics of India's approach to climate policies and negotiations, we do expect that the Indian negotiators will become even more sophisticated in arguing their case and in defending their positions, and that they will increasingly shift their focus to adaptation.

However, indirect routes to mitigation are available. General environmental consciousness in India is quite high, in part because of its cultural heritage. Mitigation actions that focus on forest and wildlife protection are likely to find traction. Two factors facilitate this process. First, environmental and forest protection are enshrined legally, and therefore the judicial system can be leveraged to protect and expand forest cover that acts as carbon sinks. Indeed, forest cover in India has been increasing since 2005 (GoI, 2013c). Second, forest protection serves adaptation goals as well as mitigation needs in sensitive regions like the

Himalayas. Thus, approaches that facilitate mitigation that are couched in environmental protection terms are gaining increasing traction, especially in the opinion-forming media. There are other encouraging factors too. Given that India imports fossil fuels, the authorities have a natural instinct to conserve fuel. Indeed, throughout India's major cities, especially New Delhi, placards exhorting citizens to use fuel economically are ubiquitous. Thus, efforts focused on energy efficiency can deliver some of the elusive win-win solutions. Finally, if the normal structures of government are unwilling or unable to tackle these issues, there is a slim but distinct possibility to appeal directly for judicial redress from India's independent and activist courts.

NOTES

1. The number of states, their names and their boundaries are constantly changing.
2. The President and the upper house of Parliament have some veto points.
3. While this chapter was being prepared, national elections in India led to the formation of a new central government, led by the Bharatiya Janata Party (BJP). The BJP government has downgraded the powers of the Planning Commission and has shifted to the Finance Ministry the responsibility for allocating funds to the central ministries and states. It is expected to officially abolish the Planning Commission.
4. Although the candidates of the IAS are selected centrally and their service conditions are set by the Union government, once recruited to the service, they are assigned to a state cadre and spend their career in state government, with periodic deputations to the centre. Generally, IAS candidates are assigned to the cadre of a state that is different from their state of origin.

REFERENCES

Agarwal, A. and S. Narain (1991), *Global Warming in an Unequal World: A Case of Environmental Colonialism*, New Delhi: Centre for Science and Environment.

Altenburg, T. and T. Engelmeier (2013), 'Boosting solar investment with limited subsidies: rent management and policy learning in India', *Energy Policy*, **59**(C), 866–74.

Atteridge, A., M.K. Shrivastava, N. Pahuja and H. Upadhyay (2012), 'Climate policy in India: what shapes international, national and state policy?', *AMBIO*, **41**(1), 68–77.

CensusInfo India (2011), 'Kitchen and type of firewood used for cooking', accessed 22 January 2015 at http://www.devinfolive.info/censusinfo dashboard/.

Chaudhary, A., C. Krishna and A. Sagar (2014), 'Policy making for renewable energy in India: lessons from wind and solar power sectors', *Climate Policy*, **15**(1), 58–87.

Columbia University (2014), 'Fossil fuel CO_2 emissions', accessed 16 January 2015 at http://www.columbia.edu/~mhs119/CO2Emissions/Emis_moreFigs/.
da Costa, A. (2013), 'OPINION: will India's elections sway climate policy?', *Worldwatch.org*, accessed 22 January 2015 at http://www.worldwatch.org/node/6072.
Dholakia, H.H., P. Purohit, S. Rao and A. Garg (2013), 'Impact of current policies on future air quality and health outcomes in Delhi, India', *Atmospheric Environment*, **75**(0), 241–8.
Dubash, N.K. (2013), 'The politics of climate change in India: narratives of equity and cobenefits', *Wiley Interdisciplinary Reviews: Climate Change*, **4**(3), 191–201.
Emission Database for Global Atmospheric Research (EDGAR) (2015), 'GHG time series 1990–2012 per capita emissions for world countries', European Commission, accessed 5 January 2015 at http://edgar.jrc.ec.europa.eu/overview.php?v=GHGts_pc1990-2012.
Gallucci, M. (2015), 'Pressure mounts on India to tackle greenhouse gas emissions ahead of 2015 climate talks in Paris', *International Business Times*, accessed 16 January 2015 at http://www.ibtimes.com/pressure-mounts-india-tackle-greenhouse-gas-emissions-ahead-2015-climate-change-talks-1782296.
Government of India (GoI) (2008), *National Action Plan on Climate Change*, Prime Minister's Council on Climate Change.
Government of India (GoI) (2011a), 'Smt. Preneet Kaur at the Post-Forum Dialogue Partners' Meeting', Ministry of External Affairs, 12 July.
Government of India (GoI) (2011b), 'Press communique', Prime Minister's Office, 12 July 2011.
Government of India (GoI) (2011c), *Strategic Plan for New and Renewable Energy for Period 2011–2017*, Ministry of New and Renewable Energy.
Government of India (GoI) (2013a), *12th Five Year Plan*, Planning Commission, New Delhi: Sage Publications.
Government of India (GoI) (2013b), *Energy Statistics 2013*, Ministry of Statistics and Programme Implementation, New Delhi: Central Statistics Office.
Government of India (GoI) (2013c), *India State of the Forest Report 2013*, Ministry of Environment and Forests, Dehradun: Forest Survey of India.
Government of India (GoI) (2014a), 'Functions and responsibilities', Government of India – Ministry of Coal website, accessed 27 April 2015 at http://www.coal.nic.in/content/functions-and-responsibilities.
Government of India (GoI) (2014b), 'Mission', Government of India – Ministry of New and Renewable Energy website, accessed 26 August 2014 at http://www.mnre.gov.in/mission-and-vision-2/mission-and-vision/mission-2/.
Government of India (GoI) (2014c), Government of India – Ministry of Petroleum & Natural Gas website, accessed 26 August 2014 at http://petroleum.nic.in/.
Government of India (GoI) (2014d), 'Responsibilities', Government of India – Ministry of Power website, accessed 26 August 2014 at http://powermin.nic.in/responsibilities.
Gupta, S. (2011), 'Jairam loses environment, Moily law in reshuffle', *The Hindu*, 13 July, accessed 17 September 2014 at http://www.thehindu.com/news/national/article2220917.ece.

International Energy Agency (IEA) (2014), 'CO2 emissions from fuel combustion – 2014 edition', Paris: International Energy Agency.

International Institute for Sustainable Development (IISD) (2012), *A Citizens' Guide to Energy Subsidies in India*, Winnipeg: International Institute for Sustainable Development.

Kallbekken, S., H. Sælen and A. Underdal (2014), *Equity and Spectrum of Mitigation Commitments in the 2015 Agreement*, TemaNord Report No. 519. Copenhagen: Nordic Council of Ministers.

Michaelowa, K. and A. Michaelowa (2012), 'India as an emerging power in international climate negotiations', *Climate Policy*, **12**(5), 575–90.

Ministry of Environment and Forests (MoEF) (2004), *India's Initial National Communication to the United Nations National Framework Convention on Climate Change*, accessed 10 January 2015 at http://unfccc.int/resource/docs/natc/indnc1.pdf.

Ministry of Environment and Forests (MoEF) (2012), *India: Second National Communication to the United Nations Framework Convention on Climate Change*, accessed 10 January 2015 at http://unfccc.int/resource/docs/natc/indnc2.pdf.

NDTV (2010), 'BJP, Left slam Jairam Ramesh on shift in climate stand', *NDTV.com*, 10 December, accessed 17 September 2014 at http://www.ndtv.com/article/india/bjp-left-slam-jairam-ramesh-on-shift-in-climate-stand-71665.

Parliament of India (2009), 'Statement by Prime Minister. Statement regarding Prime Minister's recent visits to Italy, France and Egypt', 17 July, accessed 28 April 2015 at http://www.hcidhaka.gov.in/pages.php?id=429.

Parliament of India (2010), 'The Minister of State of the Ministry of Environment and Forests made a statement regarding issues relating to Copenhagen Accord', accessed 28 April 2015 at http://indiankanoon.org/doc/546743/.

Parliament of India (2013), 'Discussion on the situation arising due to drought and hailstorms in various parts of the country', accessed 28 April 2015 at http://indiankanoon.org/doc/195671819/.

Phillips, J. and P. Newell (2013), 'The governance of clean energy in India: the Clean Development Mechanism (CDM) and domestic energy politics', *Energy Policy*, **59**(C) 654–62.

Planning Commission (2006), *Integrated Energy Policy: Report of the Expert Committee*, Government of India: New Delhi.

Rastogi, N.P. (2011), 'Winds of change: India's emerging climate strategy', *The International Spectator*, **46**(2), 127–41.

Sengupta, S. (2012), 'International climate negotiations and India's role', in N.K. Dubash (ed.), *Handbook of Climate Change and India: Development, Politics and Governance*, New Delhi: Oxford University Press, pp. 101–17.

Sharma, B. (2014), 'India sends mixed signals on climate change', *Al Jazeera*, accessed 27 October 2014 at http://www.aljazeera.com/indepth/features/2014/09/india-sends-mixed-signals-climate-change-2014926202730167480.html.

Singh, M.K. (2014), 'New poverty line: Rs 32 in villages, Rs 47 in cities', *Times of India*, 7 July, accessed 2 September 2014 at http://timesofindia.indiatimes.com/india/New-poverty-line-Rs-32-in-villages-Rs-47-in-cities/articleshow/37920441.cms.

Tankha, S. (2009), 'Building administrative capacities in developing countries: SWAT teams or beat cops?', *Public Administration Review*, **69**(6), 1028–30.
Tankha, S. and B.W. Fuller (2010), 'Getting things done: bureaucratic and entrepreneurial approaches to the practice of participatory water management reforms in country-region Brazil and place country-region India', *Water Policy*, **12**(1), 84–103.
Tankha, S. and T. Rauken (2013), *Institutional Capacities for Climate Change Adaptation in India – A Pilot Study*, Report 2013:04, Oslo: CICERO.
Tankha, S., A.B. Misal and B.W. Fuller (2010), 'Getting reforms done in inhospitable institutional environments: untying a Gordian Knot in India's power distribution sector', *Energy Policy*, **38**(11), 7121–9.
The Economist (2008), 'China, India and climate change – melting Asia', *Economist.com*, 5 June, accessed 15 January 2015 at http://www.economist.com/node/11488548.
US Energy Information Administration (EIA) (2014), 'India', accessed 5 January 2015 at http://www.eia.gov/countries/cab.cfm?fips=in.
World Bank (2013), 'India: climate change impacts', accessed 23 January 2015 at http://www.worldbank.org/en/news/feature/2013/06/19/india-climate-change-impacts.
World Bank (2014a), 'CO2 emissions (metric tons per capita)', accessed 23 January 2015 at http://data.worldbank.org/indicator/EN.ATM.CO2E.PC.
World Bank (2014b), 'Electric power consumption (kWh per capita)', accessed 23 January 2015 at http://data.worldbank.org/indicator/EG.USE.ELEC.KH.PC.
World Bank (2014c), 'Energy use (kg of oil equivalent per capita)', accessed 23 January 2015 at http://data.worldbank.org/indicator002FEG.USE.PCAP.KG.OE.
World Bank (2014d), 'GINI index', accessed 23 January 2015 at http://data.worldbank.org/indicator/SI.POV.GINI.
World Bank (2014e), 'Poverty headcount ratio at $1.25 a day (PPP) (% of population)', accessed 23 January at http://data.worldbank.org/indicator/SI.POV.DDAY.
World Bank (2015a), 'India', accessed 30 April 2015 at http://data.worldbank.org/country/india?display=default.
World Bank (2015b), 'Access to electricity (% of population)', accessed 30 April 2015 at http://data.worldbank.org/indicator/EG.ELC.ACCS.ZS.
World Bank (2015c), 'GDP per capita (current US$)', accessed 30 April 2015 at http://data.worldbank.org/indicator/NY.GDP.PCAP.CD?display=default.
World Bank (2015d), 'GDP per capita growth (annual %)', accessed 30 April 2015 at http://data.worldbank.org/indicator/NY.GDP.PCAP.KD.ZG?display=default.
World Bank (2015e), 'Population growth (annual %)', accessed 30 April 2015 at http://data.worldbank.org/indicator/SP.POP.GROW?display=default.
World Bank (2015f), 'Population density (people per sq. km of land area)', accessed 30 April 2015 at http://data.worldbank.org/indicator/EN.POP.DNST?display=default.
World Bank (2015g), 'Land area (sq. km)', accessed 30 April 2015 at http://data.worldbank.org/indicator/AG.LND.TOTL.K2?display=default.

WRI (2014), 'CAIT 2.0 WRI's climate data explorer. GHG emissions by sector', accessed 5 January at http://cait2.wri.org/wri/Country%20GHG%20Emissions?indicator[]=Energy&indicator[]=IndustrialProcesses&indicator[]=Agriculture&indicator[]=Waste&indicator[]=Land-UseChangeandForestry&indicator[]=Bunker Fuels&year[]=2011&act[]=India&chartType=pie.

Yeo, S. (2014), 'Will India's new PM Modi be a climate change champion?', *RTCC*, 19 May, accessed 27 October 2014 at http://www.rtcc.org/2014/05/19/will-indias-new-pm-modi-be-a-climate-change-champion/.

6. Japan's climate policy: post-Fukushima and beyond

Masahiko Iguchi, Alexandru Luta and Steinar Andresen

BOX 6.1 KEY FACTS ABOUT JAPAN

Population 2013	127.3 million
Population growth rate 2013	0.2% decrease from 2012, gradual decrease since 2008
Population density 2005	343 per km^2
GDP per capita 2013	38 492.1 USD
GDP growth rate 2013	1.5%
GINI index 2008	32.1
Access to electricity 2015	100% of the population
Land area	377 900 km^2
Political system	Multi-party parliamentary representative democratic constitutional monarchy
Subnational entities	47 prefectures

Sources: Japan Statistics Bureau (2014); World Bank (2014a, 2014b, 2015).

INTRODUCTION

Japan (Box 6.1) is in a unique position in international climate politics. Even though it hosted the third Conference of the Parties (COP) to the United Nations Framework Convention on Climate Change (UNFCCC) in 1997 in order to take leadership by promoting its environmental reputation, it has recently rejected participation in the second commitment period of the Kyoto Protocol. Moreover, although Japan has committed to reduce its greenhouse gases (GHG) emissions by 6 per cent under the Kyoto Protocol, its emissions have actually *increased* more than 6 per cent by 2012 from the 1990 level (see Table 6.1). Finally, despite leading the international GHG emissions reduction pledges by proposing a 25 per cent decrease from the 1990-level target in 2009, Japan's latest proposed mid-term target sets a reduction of 26 per cent from the 2013 level by 2030 (equivalent to an 18 per cent reduction from 1990 level).

What has caused this backlash? Is it Japan's high dependence on fossil fuels due to the Fukushima nuclear disaster – or are there more fundamental stumbling blocks? The following sections examine the characteristics of Japanese climate policy, and note the implications for its future policy trajectory.

Table 6.1 GHG emissions and energy data for Japan

Total GHG emissions by year	1990	2005	2012
	1261 $MtCO_2e$[a]	1350 $MtCO_2e$	1343 $MtCO_2e$
GHG emissions per capita 2012	9.34 tCO_2e[b]	10.21 tCO_2e	10.16 tCO_2e
GHG emissions by sector 2012	Industry 34.6%, transport 18.7%, commercial and other 22.6%, residential 16.9%, energy industries 7.3%		
Energy mix 2013	Fossil fuels 88.3%, renewables 10.7%, nuclear 1%		

Notes:

a. Million metric tons CO_2 equivalents.
b. Tons CO_2 equivalents.

Sources: FEPC (2014); MOE (2014); World Bank (2014a); NIES (2015).

THE DEPENDENT VARIABLE: CLIMATE CHANGE MITIGATION POLICY

This section uses overall policy shifts as the dependent variable, from the Kyoto Protocol in 1997 to the current state of Japanese climate policy after the Fukushima nuclear disaster in 2011. It shows the unstable grounds of Japan's climate policy along with frequent administrative changes.

Japan's Current Policies and Recent Trends, Mid- to Long-term

Since the adoption of the Kyoto Protocol in 1997 in which Japan made a reduction commitment of 6 per cent from the 1990 level, various mid- and long-term GHG reduction targets have been proposed, under new prime ministers between 2007 and 2015 (Table 6.2).

Table 6.2 Milestones in Japanese climate mitigation policy

Month/Year	Targets	Prime Minister (Party and Term of Appointment)
1997	6% reduction from 1990 levels under the Kyoto Protocol	Ryutaro Hashimoto (Liberal Democratic Party, hereafter LDP, 1996–98)
May 2007	50% reduction of global CO_2 emissions by 2050	Shinzo Abe (LDP, 2006–07)
January 2008	14% reduction from 2005 levels by 2020; long-term target: 60–80% from 2005 levels (no specific year set)	Yasuo Fukuda (LDP, 2007–08)
June 2009	15% reduction from 2005 levels (equals 8% compared to 1990 level) by 2020	Taro Aso (LDP, 2008–09)
September 2009	25% reduction from 1990 levels by 2020 (pledged in the Copenhagen Agreement in January 2010)	Yukio Hatoyama (Democratic Party of Japan, hereafter DPJ, 2009–10)
March 2013	Review of the 25% target by 2020	Shinzo Abe (LDP, 2012–Present)
November 2013	Proposed 3.8% reduction compared to 2005 level by 2020	Shinzo Abe (LDP, 2012–Present)
June 2015	New target of 26% reduction compared to 2013 level by 2030	Shinzo Abe (LDP, 2012–Present)

Source: Authors' files.

The first proposed target beyond 2012 was the *long-term* target of reducing global carbon dioxide (CO_2) emissions by 50 per cent by 2050, announced by the first Abe administration (LDP, 2006–07) in the 'Cool Earth 50' initiatives during the 33rd G8 Summit in 2007 (Prime Minister of Japan and His Cabinet, 2007). This target was further advanced by the next prime minister, Yasuo Fukuda (LDP, 2007–08). At the 34th G8 Summit hosted in 2008 by Japan in Toyako, Hokkaido, he announced the 'Fukuda Vision', setting Japan's long-term target to reduce GHG emissions by 60 to 80 per cent from current levels. He also noted the potential *mid-term* target of 14 per cent reduction from 2005 levels by 2020 (Prime Minister of Japan and His Cabinet, 2008).

Discussions regarding Japan's mid-term reduction target were developed during the Aso administration (LDP, 2008–09), which established the 'Mid-Term Target Committee' under the Council on the Global Warming Issue. At seven meetings held between November 2008 and April 2009, the committee discussed targets for 2020, and proposed six, ranging from a 4 per cent reduction from the 2005 level (equivalent to 4 per cent increase from the 1990 level) to a 30 per cent reduction from the 2005 level (equivalent to 25 per cent reduction from the 1990 level) (Mid-Term Target Committee, 2009). In June 2009, the administration announced a reduction of 15 per cent from the 2005 level (option 5) by 2020 (Prime Minister of Japan and His Cabinet, 2009a).

Hence, Japan's first mid-term target was proposed in 2009; however, it was not particularly ambitious, since GHG emissions had peaked in 2005, with an increase of 890 $MtCO_2e$ from 1990 (Figure 6.1). The preference for 2005 as a base year arose from industry claims Japan had already achieved high energy efficiency after experiencing the two oil crises of the 1970s. Using 1990 as the reference year put Japanese industry at a disadvantage compared to Europe, with its large potential for emissions reduction due to the EU enlargement.

A significant change came when former Prime Minister Yukio Hatoyama (DPJ, 2009–10) announced a 25 per cent reduction target (on a conditional basis, with the premise of the establishment of a fair and effective international framework where all major economies participate and agree on ambitious targets) at the Copenhagen UN Summit on Climate Change (Prime Minister of Japan and His Cabinet, 2009b). The target became Japan's pledge to the Copenhagen Accord (UNFCCC, 2010). In order to legalize this commitment, the bill of the Basic Law on Global Warming Countermeasures, which mandated GHG emissions reductions of 25 per cent by 2020 and 80 per cent by 2050, was approved by the Hatoyama Cabinet in March 2010. The proposed law included sets of policy measures that included a mandatory nationwide emissions

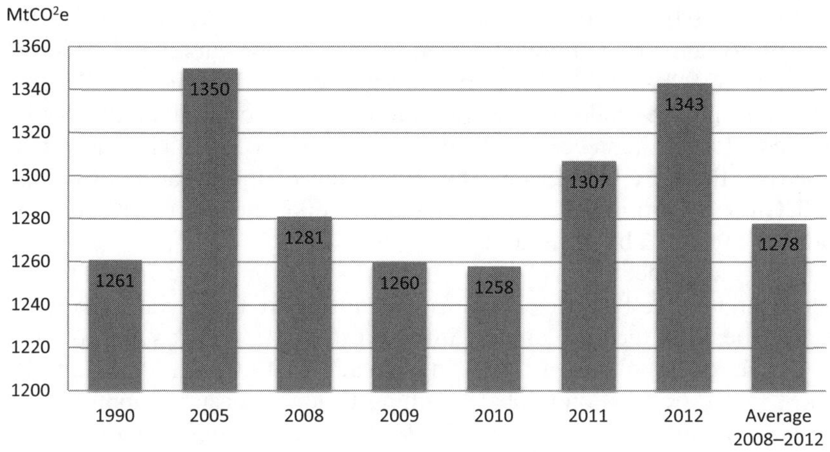

Source: Created by author based on MOE (2014).

Figure 6.1 Japan's GHG emissions 1990–2012

trading scheme (ETS), a carbon tax and a generalized feed-in tariff (FiT) for the net amount of electricity generated from all renewable sources.

However, this proposed law failed to pass the Diet, as the DPJ lost its Upper House majority in 2010. The bill was never revived, as the crisis triggered by the 2011 Great East Earthquake and the Fukushima Daiichi nuclear disaster comprehensively reset the priorities of Japanese politics. Moreover, as achieving a 25 per cent reduction target depended partially on the use of nuclear power generation, the target itself became a subject of reconsideration. It was the second Abe administration in 2013 that officially announced reconsideration of the 25 per cent reduction target (Prime Minister of Japan and His Cabinet, 2013).

In November 2013 Japan proposed a target of 3.8 per cent reduction compared to the 2005 level (equivalent to a 3.1 per cent increase from the 1990 level) by 2020. This target was subject to change according to future energy policies, and does not include electricity generation by nuclear power. Then, in June 2015, Japan announced a reduction target of 26 per cent from the 2013 level by 2030.

Main Domestic Measures

Many attempts have been made to create a general, non-legally binding framework on climate change, including an Action Plan to Arrest Global

Warming (1990), the Basic Policy on Measures to Tackle Global Warming (1999), and the Outline for Promotion of Efforts to Prevent Global Warming (1998, 2002). The first framework for achieving Japan's emissions pledge under the Kyoto Protocol was drafted in the Kyoto Protocol Target Achievement Plan (KTAP) in 2005, which was established by the Law Concerning the Promotion of the Measures to Cope with Global Warming. Partially revised in 2006, the drafted KTAP was adopted by the Cabinet in 2008.

These three documents do not contain specific policy instruments, but merely enumerate the roles that various entities are to play and the types of activities that they are already involved in under pre-existing arrangements. Tangible policies for reducing Japan's GHG were initiated by the bill of the Basic Law on Global Warming Countermeasures, approved by the Hatoyama Cabinet in March 2010. Although the legalized emission reduction targets of 25 per cent by 2020 and 80 per cent by 2050 had not materialized at the time of the resignation of the Hatoyama Cabinet, two out of three core climate policies of the bill were further facilitated. Those were the FiT for Renewable Energy, implemented from July 2012 onwards, and the Tax for Climate Change Mitigation (carbon tax) enacted in October 2012.

Japan's FiT was established under the Act of Purchase of Renewable Energy Sourced Electric Utilities in 2012. It requires electric utilities to purchase electricity generated from renewable energy sources, including solar PV, wind power, hydraulic power, geothermal and biomass, based on a fixed time period with fixed price. Although its effect is yet to be seen, the share of renewable energy (geothermal and other new energy) increased slightly from 1.6 per cent in 2012 to 2.2 per cent in 2013 (FEPC, 2014).

The carbon tax imposes JPY 289 per ton of CO_2 for all fossil fuels; tax rates will be raised in three stages over three-and-a-half years (2012–16). This is expected to generate revenue of JPY 262.3 billion in total, to be used for the introduction of renewable energy and further enhancements of energy-saving measures (MOE, 2012a).

The third measure, introduction of nationwide ETS, has not materialized. Instead, the Tokyo metropolitan government introduced a mandatory cap-and-trade programme, with the target of 6 per cent GHG emissions reduction between 2010 and 2014 and a prospective target of 17 per cent reduction between 2015 and 2019 (Tokyo Metropolitan Government, 2010). Already in 2012, the programme achieved a reduction of 13 per cent by March 2011 and 23 per cent by March 2012, due to significant energy savings after the power crisis that ensued with the Great East Earthquake (Tokyo Metropolitan Government, 2013).

Besides these economic instruments, Japan has developed several measures for reducing its GHG emissions, including the enhancement of energy efficiency through top-runner method,[1] a consumer-oriented approach such as the 'Cool Biz' campaign,[2] and the Voluntary Action Plan (VAP)[3] by the high-level Japan Business Federation (Keidanren). However, in the absence of a mandatory emissions reduction target and a nationwide ETS, these policies remain rather complementary, as Japan's GHG emissions rose to 1343 MtCO_2e in 2012, and its average emissions between 2008–12 exceeded 18 MtCO_2e from the 1990 level (equal to a 1.4 per cent increase). Forest sink and the Kyoto Protocol's Flexible Mechanisms – the Clean Development Mechanism (CDM) and Joint Implementation – have been Japan's major instruments for achieving its Kyoto commitment of 6 per cent reduction. Statistics show that Japan reduced its emissions by 8.4 per cent from the 1990 level by using 3.9 per cent from forest sink, and 5.9 per cent from Kyoto mechanisms (MOE, 2014).

Japan's Role in International Climate Cooperation

Despite the absence of effective domestic climate measures, Japan, led by the Ministry of Foreign Affairs (MOFA), was keen to promote its environmental reputation abroad so as to exert leadership during the Kyoto Protocol negotiation in 1997 (Hattori, 2007). The literature points out that the USA has often influenced Japan's foreign policy (Inoguchi, 1993; Iida, 1999; Ikenberry and Inoguchi, 2003). However, Japan did ratify the Kyoto Protocol in 2001, despite the US withdrawal.

On the other hand, while Japan has committed to GHG emissions reductions under the Kyoto Protocol, it has also joined legally non-binding, technology-oriented agreements such as the 2005 US-led Asia-Pacific Partnership on Clean Development and Climate (APP), held up as an alternative to the legally binding framework under the Protocol (Christoff and Eckersley, 2007; Lawrence, 2007; McGee and Taplin, 2009). One reason why Japan joined such an agreement was to play an intermediary role after the US rejection of the Kyoto Protocol, and hence to mediate between an environmentally passive USA and an active Europe (Miyaoka, 2004; van Asselt et al., 2009). More importantly, Japan was keen to promote a 'sectoral approach' – pursuing GHG mitigation based on different sectors internationally, thereby relieving pressures on industry competitiveness – to be integrated into the post-2012 framework, as long promoted by Keidanren (Keidanren, 2011).

At the 2009 Copenhagen summit, Japan (together with Russia and Canada) argued forcefully for the creation of an ambitious new treaty that would be binding for all major emitters in developed and developing countries alike – instead of favouring the second commitment period of the Kyoto Protocol (Dimitrov, 2010). Japan continued this position in subsequent negotiations, and expressed its opposition to participating in the second commitment period of the Kyoto Protocol, reasoning that it would 'lead to solidifying the unfair and ineffective framework beyond 2012' (MOFA, 2010).

These interests gradually led Japan to initiate a new climate change institutional package, the East Asia Low-Carbon Growth Partnership (LCGP), proposed at the East Asian Summit of Foreign Ministers' Consultation in July 2011. As Japan is no longer eligible for the CDM, because of its decision not to participate in the Kyoto Protocol's second commitment period, the government has steadily been exploring other mechanisms, outside the UNFCCC, which could enable the flow of technologies from Japan to other countries, with the possibility that some offsets might go in the reverse direction.

As the APP concluded its work in April 2011 and Japan had officially announced that it would not participate in the second Kyoto period, Japan's two-track approach between the Kyoto Protocol and the APP ended with the creation of the LCGP (Oh and Matsuoka, 2013). Like the APP, the LCGP is a non-UNFCCC track that aims to promote 'models for low-carbon growth which serve economic growth and challenges against global warming by sharing member countries' efforts, experience and environmental technology through promoting regional cooperation' (MOFA, 2014). Although it has some characteristics of the APP, its impact on the forthcoming negotiations will need to be a subject for future research.

As regards financing, Japan has played an important role at the UNFCCC negotiations. For instance, it committed to provide USD 11 billion in state funding for 2010–12 at the Copenhagen negotiations, which was the highest among countries such as the EU (10.7 billion), Norway (10.5 billion), Russia (200 million) and even the USA (3.7 billion). Furthermore, it donated USD 15 billion from public and private finance through 'fast-start finance' for developing countries up to 2012, in order to assist developing countries with mitigation and adaptation measures.

EXPLAINING POLICY TRAJECTORIES

Fundamental Parameters

Material parameters: economic recession and industry competitiveness

Thanks to improvements made in the 1980s, Japan ranked highest in the world in energy efficiency for much of the ensuing two decades. However, that also meant that the marginal costs of additional mitigation in the industrial and energy sector were much higher there. Even today, Japan faces much higher costs than many of its neighbours and economic competitors, like the USA, China and India.

These aspects weigh heavily when it comes to selecting policies to tackle climate change, especially given the slow growth rates after the burst of the bubble economy that had driven the Japanese boom in the 1980s. What resulted was a balance-sheet recession, where, despite the near-zero interest rates offered by the Bank of Japan, companies opted to deleverage their positions and save instead of investing. With domestic consumption depressed as a result of deflation, government expenditure and net exports have been the main drivers of economic growth.

Japan's export-driven strategy, with trade barriers sheltering the domestic economy from competition from foreign goods, can be traced to the aftermath of World War II. Although the development of the world trade regime, from the General Agreement on Tariff and Trade to the World Trade Organization, has meant that Japan has had to peel back these barriers and progressively lower its tariffs, the importance of exports to the economy is firmly rooted in its economic policy, especially since the possibility of expanding other elements of GDP is uncertain. Thus, maintaining the competitiveness of Japanese industries remains a factor of extreme concern, as the exporting manufacturing industries (automotive, consumer electronics, semiconductors and so forth) sector accounts for more than 20 per cent of GDP.

This has obstructed the development of Japanese climate policy, domestically and internationally, because of the continued perception that meddling with fundamental areas such as energy or industrial policy would yield ripple effects ruinous to the Japanese economy – especially if other major emitters had not pledged similar mitigation efforts.

Furthermore, energy efficiency improvements during the 1980s led to significant technological advances in Japan in various fields, including steel manufacturing and electricity generation. This technological progress may have been a central reason for Japan's choice of a sectoral approach under the post-2012 framework, prior to the 2009 Copenhagen climate conference. Such an approach would have enabled Japan to export its technologies, while reaping the mitigation benefits by counting the emission cuts stemming from deployment of those technologies abroad against its own national GHG account.

Material parameters: energy and the Fukushima Daiichi nuclear disaster in 2011

As a resource-poor country reliant on foreign imports to meet its energy needs, Japan has pursued climate policies heavily influenced by its energy policies. Earlier reliance on Middle Eastern oil led to two oil crises in the 1970s, to which Japan's Ministry of International Trade and Industry (reorganized into the Ministry of Economy, Trade and Industry, METI, in 2001) promptly reacted by diversifying, to reduce the country's dependence on oil. That meant greater use of coal, as well as gas and nuclear power, and approaches like increased energy efficiency and use of non-conventional energy sources. The Act on the Rational Use of Energy (1979) led to increased energy efficiency in the Japanese economy during the 1980s.

Energy security has long been an important priority in Japan. Ever since the 1950s, the emphasis has been on nuclear energy, with the Atomic Energy Basic Act (1955) and the Act on the Regulation of Nuclear Source Material, Nuclear Fuel Material Reactors (1957). The METI has played a leading role in the promotion and regulation of commercial nuclear power plants (Iguchi and Koga, 2015). The call for the steady expansion of the role of nuclear power was emphasized by the Basic Energy Plans (2003), revised in 2007 and 2010, specifying that 14 new nuclear reactors were to be added, increasing Japan's dependence on nuclear power to about 50 per cent by 2030 (METI, 2010).

However, on 11 March 2011 came the Great East Earthquake, triggering the triple meltdown at the Fukushima Daiichi nuclear power plant – and dealing a severe blow to the image of nuclear power. The course previously staked out would be difficult to pursue. Concerns about the use of nuclear power grew among the Japanese public, and gradually led to the anti-nuclear and denuclearization movements. The Kan administration (DPJ, 2010–11) reacted by urging a review and re-evaluation of the country's nuclear policy, and establishing the Energy and Environment Council in collaboration with the METI. The Council produced three scenarios to reduce the share of nuclear generation by 2030: zero scenario, 15 per cent scenario and 20–25 per cent scenario (Table 6.3).

Table 6.3 Scenarios to reduce share of nuclear power generation by 2030 (%)

	Zero Scenario	15% Scenario	20–25% Scenario
Share of nuclear generation	0	15	20–25
Share of renewable energy	35	30	25–30
Share of fossil fuels	65	55	50

Source: MOE (2012b).

The Japanese public clearly favoured the zero scenario, with 90 per cent of those surveyed choosing that option (National Policy Unit, 2012). Nuclear power has lost public trust, but there are high expectations to renewable energy as an alternative energy source. As a result, the government set about reviewing its position on nuclear energy, and launched the Act on Special Measures concerning the Procurement of Renewable Energy by Operators of Electric Utilities in August 2011. The Act mandated FiTs for renewable energy: electric utility companies are required to purchase renewable energy, which includes electricity generated from solar, wind, hydro, geothermal and biomass sources. However, although Japan introduced a generous FiT system in 2012, renewable energy poses several challenges regarding grid integration, and also starts from a very low base. Since the second Abe administration took power in 2012, the government has been considering a future energy mix from scratch with an adoption of the fourth Strategic Energy Plan in April 2014 (METI, 2014b).

Another important aspect of Japan's post-Fukushima energy policy is the increased dependence on fossil fuel (Figure 6.2). With political constraints preventing utilities from restarting nuclear reactors, the shortfall in generation capacity must be met with fossil fuels, which drove emissions to 14 per cent above the 1990 base year in fiscal year 2012. This poses serious challenges to the formulation of a Japanese target for mid-term mitigation efforts. Whereas Keidanren's Voluntary Action Plan (VAP) had been keeping emissions from the industrial and energy sectors under control,[4] this is no longer possible due to the sudden reliance on fossil fuel imports. Furthermore, emissions stemming from increased energy use in the household, commercial and transportation sectors are soaring. This places a great burden on the government to develop a more forceful policy package to tackle the evolution of Japan's future emissions trajectory.

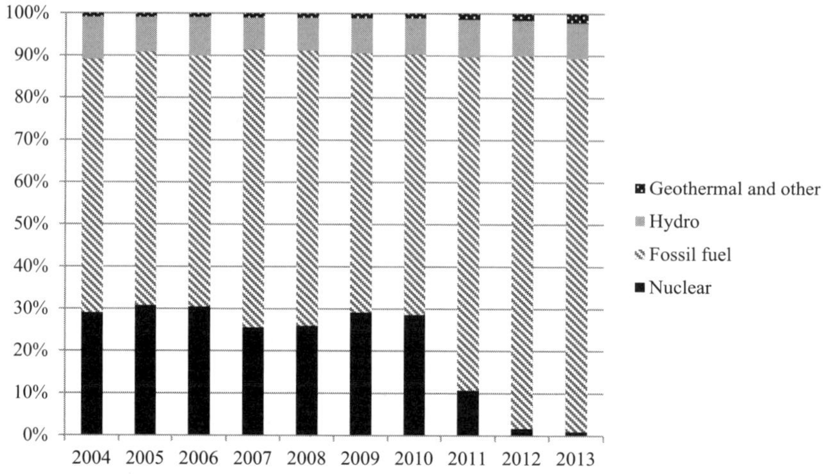

Source: FEPC (2014).

Figure 6.2 Japan's energy mix for electric generation (2004–13)

Political parameters: iron triangle, bureaucratic politics and big business

Japan is a multi-party, parliamentary representative democratic constitutional monarchy, with the emperor as symbolic head of state, and the prime minister as head of the government, with the Cabinet as an executive branch. There is no fixed term for the prime minister, but he or she must resign if the House of Representatives adopts a motion of no confidence or defeats a vote of confidence. The legislature, the Diet, consists of the House of Representatives (the lower house with 475 members, elected for four-year terms) and the House of Councillors (the upper house with 242 members, elected for six-year terms). Judicial power is vested in the Supreme Court and lower courts.

Historically, the term 'iron triangle' has been used to describe Japanese politics (Samuels, 1987; Pempel, 1998). This means that the decision-making process has been dominated by the triangle of the LDP, bureaucrats and big business supporting the political dominance of the LDP. Furthermore, 'bureaucratic politics' have been seen as the central driver of decision-making (Johnson, 1982). This also applies to climate policy-making, where not only the Ministry of the Environment (MOE), but also the METI has been actively engaged. The MOFA provides vital support in the arena of international negotiations. This jurisdictional conflict arising over an issue that crosses the boundaries between ministries and agencies is particularly difficult to resolve. While climate change may be considered an environmental matter, the METI has considerable influence on climate policy, arguing for the promotion of competitiveness of Japanese industry. By contrast, the MOE is relatively young, upgraded to a ministry only in 2001, with limited financial and human resources. Japanese climate policy-making has been described as fragmented, because of the conflicting perspectives involved (Takeuchi, 1998; Kawashima, 2000; Oshitani, 2006; Tiberghien and Schreurs, 2007).

Importantly, Japan's climate policy has also been affected by powerful industrial influences (Fisher, 2004). There is a close relationship between METI and business associations, which have generally managed to avoid direct regulations through their VAP. Business organizations, in alliance with the METI, have successfully resisted attempts by politicians and other ministries, chiefly the MOE, to introduce other kinds of policies by framing the discussion in terms of international competitiveness. Already by 2004, the VAP was institutionalized as the main instrument for emissions control under the KTAP of the Japanese government (Watanabe, 2011). The roles of these actors as suppliers of climate change policy are further explored in the next section.

Driving Forces for Policy Change

Climate policy: supply side

As noted, suppliers of climate policy (political parties and bureaucracy) and big business have close relationships in decision-making. Although the same types of actors (various government bureaucracies, political parties, business associations and so forth) can act as policy suppliers, as is common in other countries, Japanese climate policy has remained relatively conservative.

This is most evident in the dominance of the LDP since 1955. The party was continuously in power until 2009, except for an eight-month interlude in 1993–94, when the country was ruled by a coalition of eight smaller parties. The 2009 election marked historic leadership for the DPJ, which emerged with 308 seats in the House of Representatives with 109 seats in the House of Councillors. The LDP regained dominance with the 2012 House of Representative election (294 seats) and the House of Councillors election (115 seats) in 2013.

The LDP has shown a rather reluctant commitment to climate policy, failing to produce any flagship climate-related policy instruments such as the introduction of nationwide ETS. Instead, faced with the expense of climate mitigation, the LDP, attuned to the needs of business, allowed the industrial and energy sectors to determine their own action plan for domestic climate action. In 2009, in preparation for the post-2012 negotiations, the LDP offered an initial mid-term pledge of only an 8 per cent reduction target, barely two percentage points less than Japan's initial Kyoto target. Thus, the LDP has played a minimal role in developing climate policy: it is industry, working together with METI, that has served as the de facto supplier of Japanese climate policy.

In contrast, the DPJ, which marketed itself as the major alternative to the establishment LDP in 1998, produced an ambitious mid-term target of 25 per cent reduction ahead of the Copenhagen conference. In preparation for the 2009 elections, the DPJ prepared a comprehensive policy manifesto that included three explicit policy initiatives on climate: the introduction of a mandatory nationwide ETS that would place a hard cap on absolute emissions; a carbon tax; and a generalized FiT to finance the total generation of electricity from all renewable energy sources. However, once in power, the DPJ's interest in climate initiatives ebbed rapidly, as it failed repeatedly to pass a basic law on global warming that could serve as a coordinating framework for its climate initiatives.

The frames dominating climate policy remain those set according to the terms of Japanese industry and business, which generally avoided direct regulations through their VAP. This was evident in discussions on the introduction of a nationwide ETS. Industry was strongly opposed, arguing that the VAP was enough; further, that imposing a mandatory nationwide ETS would harm the competitiveness of Japanese industry, leading business to leave Japan for other locations – in effect causing carbon leakage.

Table 6.4 outlines the reduction commitments of each industry association. A study conducted for Keidanren in September 2010 showed that 61 companies out of 64 surveyed opposed the introduction of ETS, and none favoured its introduction (Keidanren, 2010). As a result, instead of a mandatory system, a voluntary ETS was introduced by the Ministry of the Environment and the Ministry of the Economy, Trade and Industry.

Table 6.4 Industries' reduction commitments under the Keidanren commitment to a low-carbon society

Industry Association	Reference Year	Target Indicator	Reduction Target
Federation of Electric Power Companies of Japan	–	CO_2 emissions intensity	0.33 kg CO_2/kWh[a]
Japan Iron and Steel Federation	2005	CO_2 reduction from business-as-usual	5 million tonnes CO_2
Japan Chemical Industry Association	2005	CO_2 reduction from business-as-usual	1.5 million tonnes CO_2
Japan Paper Association	2005	CO_2 reduction from business-as-usual	1.39 million tonnes CO_2
Electronics and Electronic Appliances	2012	Energy consumption rate	More than 7.73%
Japan Cement Association	2010	Energy reduction from business-as-usual	0.056 million kl
Japan Automobile Manufacturers' Association	1990	CO_2 emissions	28%
Petroleum Association of Japan	2010	Energy conservation	0.53 million kl
Japan Gas Association	1990	CO_2 reduction from business-as-usual; energy consumption rate	89%; 85%

Note: a. Under reconsideration after the 2011 Great East Earthquake.

Source: METI (2014a).

Climate policy: demand side

The pursuit of more concrete climate policy in Japan has been demanded by many actors, ranging from the international community, to civil society and businesses seeking to challenge the establishment of Keidanren. While the international community could be powerful, demanding specific policy from the Japanese government is a matter to be settled by domestic stakeholders. In contrast to Europe, where calls for progressive action on climate were channelled towards governments by means of civil society movements, often through green parties, Japan faced obstacles that prevented the emergence of a strong civil society or a Japanese green party to mainstream climate change as an issue on the public agenda (Foljanty-Jost, 2005).

Japanese civil society has developed along quite different dimensions than in Western Europe or North America. Although the Japanese public is increasingly concerned with climate change and clearly supports more proactive measures,[5] the law places heavy administrative burdens on prospective environmental non-governmental organizations (ENGOs) as regards budgeting, organization and planning, all of which must be reported to various government offices. These constraints make it very difficult for a genuine national movement concerned with problems on the scale of climate change to emerge. In the late 1960s and early 1970s, there were national citizen movements concerned with the environment because of the severe pollution caused by industrial expansion, which led to the passage of a 1970 Law for the Punishment of Environmental Pollution on Crimes relating to Human Health. However, once these goals were achieved, the protests lost momentum, and it remained difficult for ENGOs to become institutionalized in decision-making processes (Kawato et al., 2008).

Another potential demander of a more stringent climate policy could be the businesses seeking to challenge the established domination of Keidanren. As noted in the literature on the role of business in global environmental governance, while business attempts to block any improvement in environmental regulations through lobbying activities (Newell and Paterson, 1998; Levy and Newell, 2005), it could support higher environmental regulations in order to create a global or regional level playing field (Vogel, 2003; Falkner, 2008). In other words, any business that stood to benefit from more stringent climate measures could be a challenger to the vested interest structure.

One such potential business association is the Japan Association of Corporate Executives (hereafter Keizai Doyukai), founded in 1946 by 83 business leaders. In contrast to Keidanren, members of Keizai Doyukai participate in this association as individuals rather than through corporate membership, an arrangement that makes its position more free and open to different points of view. While acknowledging the VAP of Keidanren, Keizai Doyukai has expressed doubts as to the efficacy of such actions, and has supported the introduction of nationwide ETS (Keizai Doyukai, 2005). However, its access to the policy-making process is still limited (Itsumi, 2011). For instance, six members of the MOE's Central Environmental Council are from Keidanren, whereas only two are from Keizai Doyukai.

That newer industries have not been able to challenge the common position developed in Keidanren has kept policy initiatives that would prove costly to established industries from being implemented, but has also kept most innovative forms of climate policy instruments from emerging. As long as there is not a focused and articulate public, or a more aggressive split within the business community, the energy-intensive industries will continue to be the most effective demanders, able to frame and implement policies on their own terms.

Supply–demand relationship

Keidanren and the METI, together with the LDP, can be said to be the 'veto players' for more progressive climate change policies. Vested interests in the country's political structure obstructs long-term change towards a low-carbon society in Japan. The DPJ, once expected to act as the new climate-challenger party to the LDP, has now softened its stance on the 25 per cent reduction target. To sum up, although there are demands for ambitious climate policies in Japan, the suppliers who could fulfil such demands tend to heed industry and not civil society. For more progressive changes to occur and to become institutionally embedded, more voices of civil society as well as new challenger businesses must arise.

LIKELY POLICY TRAJECTORY

Japan's climate policy has experienced significant changes in recent years. Frequent shifts of governments between 2007 and 2013 have brought about various emissions reduction pledges. However, only the brief DPJ rule had a truly ambitious policy, but its most ambitious parts were never implemented. A major reason was the 2011 Fukushima Daiichi nuclear disaster, which altered the priorities of Japanese politics. As a result, the LDP in 2013 proposed a target that was very modest, in comparison to other states as well as to previous pledges. However, here it should be borne in mind that the costs of reducing emissions in Japan are particularly high because of its high energy efficiency score. Japan has relied primarily on voluntary policies, some of which have been quite effective in limiting emissions from business and industry. Moreover, it has recently introduced a CO_2 tax and a more aggressive policy on new renewables, although the latter starts from a very low level. In any case, we can note some promising signs in Japan's overall rather cautious climate policies.

Still, the actors demanding a more forceful climate policy are weak and dispersed, whereas the supply-side veto players are strong and concentrated. The 'iron triangle' of the LDP, bureaucrats and business associations, as well as the dominant position of METI in climate policy, has made industry and big business the de facto suppliers of climate policy. There are few indications that this is about to change in the future, and none of these actors want an ambitious climate policy. However, as they favour the reintroduction of nuclear energy this may contribute to a relative reduction of emissions over time. Although the public opposes Japan's reliance on nuclear power, the LDP tends to listen to the business community, which has warned that electricity prices will rise if the policy shift from nuclear to more expensive imported fossil fuels continues. They argue this will also harm Japan's overall competitiveness, a concern that decision-makers generally heed. Furthermore, industry has argued that renewable energy cannot expand quickly, safely and cheaply enough to make up for the shortfall from nuclear power. Thus, if emissions are to be reduced, nuclear power will probably play a more important role than will renewable energy.

Internationally Japan is likely to continue its financial generosity within the UN track. It will also continue to insist on a regime that will not put Japan at a disadvantage in relation to major economic competitors, whether in the North or in the South. Japan will also continue its participation in more exclusive forums outside the UN, emphasizing the importance of the sector approach as well as technological development.

NOTES

1. This method sets the 'top runner' (the most efficient model), and obligates the rest of the models to follow the top runner within a given time, triggering a race to the top. See Iguchi (2015) for how Japanese car manufacturers have achieved high fuel efficiency with this approach.
2. This campaign was advocated by the Ministry of the Environment to help reduce electricity consumption by limiting the use of air conditioning, such as setting thermostats at 28°C and encouraging employees to wear short-sleeved shirts without jackets and ties in the summer.
3. The VAP was originally set up in June 1997, prior to Kyoto Protocol negotiations at the COP 3. This peak body of Japan's major industrial associations had pledged to keep CO_2 emissions from industrial and energy conservation sectors below 1990 levels from 2008 to 2012.
4. According to Keidanren's 2013 follow-up report, industrial and energy conservation sectors managed to reduce emissions 12.1 per cent from the 1990 level between 2008 to 2012 (Keidanren, 2013a). In December Keidanren established a plan, 'Keidanren's Commitment to a Low Carbon Society' for the year 2020. This covers more than 86 per cent of emissions from industrial and energy conservation sectors, setting reduction commitment targets per industry, as well as estimating the reduction potential available from the development of innovative technologies (Keidanren, 2013b; METI, 2014a).

5. For instance, in an opinion poll conducted in 2007 by the government, 64.6 per cent answered that GHG reduction efforts by industry should be more proactive, while only 6.9 per cent answered that they should be less. Furthermore, 61.8 per cent opined that the government should take the lead in facilitating policy instead of depending on voluntary measures (Government of Japan, 2007).

REFERENCES

Christoff, P. and R. Eckersley (2007), 'The Kyoto Protocol and the Asia Pacific Partnership on Clean Development and Climate', in T. Bonyhady and P. Christoff (eds), *Climate Law in Australia*, Sydney: Federation Press, pp. 32–45.

Dimitrov, R.S. (2010), 'Inside UN climate change negotiations: the Copenhagen Conference', *Review of Policy Research*, **27**(6), 795–821.

Falkner, R. (2008), *Business Power and Conflict in International Environmental Politics*, Basingstoke, UK: Palgrave Macmillan.

Federation of Electric Power Companies of Japan (FEPC) (2014), 'Denryoku-betsu hatsuden denryokuryou kouseihi' [Electric power generation mix in percentages] (in Japanese), accessed 6 November 2014 at http://www.fepc.or.jp/about_us/pr/pdf/kaiken_s1_20140523.pdf.

Fisher, D. (2004), *National Governance and the Global Climate Change Regime*, Lanham, MD: Rowman & Littlefield.

Foljanty-Jost, G. (2005), 'NGOs in environmental networks in Germany and Japan: the question of power and influence', *Social Science Japan Journal*, **8**(1), 103–17.

Government of Japan (2007), 'Ondankataisaku ni kansuru yoronchosa' [A survey on climate change issues] (in Japanese), accessed 6 November 2014 at http://survey.gov-online.go.jp/h19/h19-globalwarming/index.html.

Hattori, T. (2007), 'The rise of Japanese climate policy: balancing the norms economic growth, energy efficiency, international contribution and environmental protection', in M.E. Pettenger (ed.), *The Social Construction of Climate Change*, Aldershot, UK: Ashgate, pp. 75–97.

Iguchi, M. (2015), *Divergence and Convergence of Automobile Fuel Economy Regulations: A Comparative Analysis of EU, Japan and the US*, Dordrecht: Springer.

Iguchi, M. and M. Koga (2015), 'The state of environment and energy governance in Japan', in S. Mukherjee and D. Chakraborty (eds), *Environmental Challenges and Governance: Diverse Perspectives from Asia*, London: Routledge, pp. 219–34.

Iida, K. (1999), *International Monetary Cooperation Among the United States, Japan, and Germany*, Berlin: Springer.

Ikenberry, G.J. and T. Inoguchi (2003), *Reinventing the Alliance: U.S.–Japan Security Partnership in an Era of Change*, Basingstoke, UK: Palgrave Macmillan.

Inoguchi, T. (1993), *Japan's Foreign Policy in an Era of Global Change*, London: Continuum International Publishing.

Itsumi, T. (2011), 'Political economy of climate change: the role of business interest representation in the institutionalization of emission trading systems in the EU and Japan', *International Relations*, **166**, 99–113.
Japan Statistics Bureau (2014), 'Monthly report – May 1 2014', accessed 6 November 2014 at http://www.stat.go.jp/english/data/jinsui/tsuki/index.htm.
Johnson, C. (1982), *MITI and the Japanese Miracle: The Growth of Industrial Policy, 1923–1975*, Stanford, CA: Stanford University Press.
Kawashima, Y. (2000), 'Japan's decision-making about climate change problems: comparative study of decisions in 1990 and in 1997', *Environmental Economics and Policy Studies*, **3**(1), 29–57.
Kawato, Y., R. Pekkanen and H. Yamamoto (2008), 'State and civil society in Japan', in A. Gaunder (ed.), *The Routledge Handbook of Japanese Politics*, London: Routledge, p. 126.
Keidanren (2010), 'A result of a survey conducted with Japanese companies on an introduction of a nationwide emission trading scheme' (in Japanese), accessed 6 November 2014 at https://www.keidanren.or.jp/policy/2010/081.pdf.
Keidanren (2011), 'Policy proposal: towards a new international framework for the post-Kyoto Protocol', Tokyo: Keidanren, accessed 6 November 2014 at http://www.keidanren.or.jp/english/policy/2011/087/index.html.
Keidanren (2013a), *Results of Fiscal 2013 Follow-up to the Voluntary Action Plan on the Environment (Summary)*, accessed 6 November 2014 at http://www.keidanren.or.jp/en/policy/2013/101.pdf.
Keidanren (2013b), *Keidanren's Commitment to Low-carbon Society*, accessed 29 April 2015 at http://www.keidanren.or.jp/en/policy/2013/003_commitment.pdf.
Keizai Doyukai (2005), 'Kyoto Giteisho Mokuhyotasseikeikaku ni taisuru iken' [Opinion on the plan to achieve Kyoto target] (in Japanese), Tokyo: Keizai Doyukai.
Lawrence, P. (2007), 'The Asia-Pacific Partnership on Clean Development and Climate (AP6): a distraction to the Kyoto process or a viable alternative?', *Asia Pacific Journal of Environmental Law*, **10**(3/4), 183–209.
Levy, D. and P. Newell (2005), *The Business of Global Environmental Governance*, Cambridge, MA: MIT Press.
McGee, J. and R. Taplin (2009), 'The role of the Asia-Pacific Partnership in discursive contestation of the international climate regime', *International Environmental Agreements: Politics, Law and Economics*, **9**(3), 319–36.
METI (2010), *The Strategic Energy Plan of Japan: Meeting Global Challenges and Securing Energy Futures (Revised in June 2010) (Summary)*, accessed 6 November 2014 at http://www.meti.go.jp/english/press/data/pdf/20100618_08a.pdf.
METI (2014a), 'Sangyokai no jishutorikumi nit suite' [About the Voluntary Action Plan] (in Japanese), Tokyo: METI.
METI (2014b), *Strategic Energy Plan*, accessed 6 November 2014 at http://www.enecho.meti.go.jp/en/category/others/basic_plan/pdf/4th_strategic_energy_plan.pdf.
Mid-Term Target Committee (2009), 'Chikyuondanka mokuhyo no sentakushi' [Options for mid-term targets] (in Japanese), Tokyo: Government of Japan.

Miyaoka, I. (2004), 'Japan's conciliation with the United States in climate change negotiations', *International Relations of the Asia-Pacific*, **4**, 73–96.
MOE (2012a), *Details on the Carbon Tax*, Tokyo, Japan: Ministry of the Environment (MOE).
MOE (2012b), 'Enerugi, Kankyo ni kansuru sentakushi' [Options for energy and environmental policy] (in Japanese), Tokyo: Ministry of the Environment (MOE).
MOE (2014), 'Nihon no Onshitsu kouka gas haishutsuryo' [Japan's GHG emissions] (in Japanese), accessed 6 November 2014 at http://www.env.go.jp/earth/ondanka/ghg/2012_gaiyo.pdf.
MOFA (2010), 'Japan's position regarding the Kyoto Protocol', accessed 6 November 2014 at http://www.mofa.go.jp/policy/environment/warm/cop/kp_pos_1012.html.
MOFA (2014), 'East Asia Low-Carbon Growth Partnership dialogue', accessed 6 November 2014 at http://www.mofa.go.jp/policy/environment/warm/cop/ealcgpd_1204/index.html.
National Institute for Environmental Studies (NIES) (2015), *National Greenhouse Gas Inventory Report of Japan*, Tsukuba: NIES.
National Policy Unit (2012), 'Public comment no torimatome kekka ni tsuite' [The result of public opinion] (in Japanese), accessed 6 November 2014 at http://www.cas.go.jp/jp/seisaku/npu/policy09/pdf/20120827/shiryo2-2-1.pdf.
Newell, P. and M. Paterson (1998), 'A climate for business: global warming, the state and capital', *Review of International Political Economy*, **5**(4), 679–703.
Oh, C. and S. Matsuoka (2013), 'The position of the Low-Carbon Growth Partnership (LCGP): at the end of Japan's navigation between the Kyoto Protocol and the APP', *International Environmental Agreements*, accessed 6 November 2014 at http://link.springer.com/article/10.1007%2Fs10784-013-9219-0.
Oshitani, S. (2006), *Global Warming Policy in Japan and Britain: Interactions between Institutions and Issue Characteristics*, Manchester, UK: Manchester University Press.
Pempel, T.J. (1998), *Regime Shift: Comparative Dynamics of the Japanese Economy*, Ithaca, NY: Cornell University Press.
Prime Minister of Japan and His Cabinet (2007), 'Invitation to "Cool Earth 50" – 3 Proposals, 3 Principles', accessed 6 November 2014 at http://japan.kantei.go.jp/abespeech/2007/05/24speech_e.html.
Prime Minister of Japan and His Cabinet (2008), 'Teitanso Nihon wo mezashite' [Towards a low-carbon society for Japan] (in Japanese), accessed 6 November 2014 at http://www.kantei.go.jp/jp/hukudaspeech/2008/06/09speech.html.
Prime Minister of Japan and His Cabinet (2009a), 'Mirai wo sukutta sedai ni narou' [To be a generation to save the future] (in Japanese), accessed 6 November 2014 at http://www.kantei.go.jp/jp/asospeech/2009/06/10kaiken.html.
Prime Minister of Japan and His Cabinet (2009b), 'Statement by Prime Minister Yukio Hatoyama at the United Nations Summit on Climate Change', accessed 6 November 2014 at http://japan.kantei.go.jp/hatoyama/statement/200909/ehat_0922_e.html.

Prime Minister of Japan and His Cabinet (2013), 'Dai ikkai sangyou kyousouryoku no giron wo fumaeta toumen no seisaku taiou ni tsuite' [Policy measures based on the first meeting on competitiveness of the industry] (in Japanese), accessed 6 November 2014 at http://www.kantei.go.jp/jp/singi/keizaisaisei/dai3/siryou03.pdf.

Samuels, R. (1987), *The Business of the Japanese State: Energy Markets in Comparative and Historical Perspective*, Ithaca, NY: Cornell University Press.

Takeuchi, K. (1998), *Chikyu ondanka no seiji-gaku* [Politics on Global Warming], Tokyo: Asahi-Sensho.

Tiberghien, Y. and M.A. Schreurs (2007), 'High Noon in Japan: embedded symbolism and post-2001 Kyoto Protocol politics', *Global Environmental Politics*, **7**(4), 70–91.

Tokyo Metropolitan Government (2010), *Tokyo Cap-and-Trade Program: Japan's First Mandatory Emission Trading Scheme*, Tokyo: Tokyo Metropolitan Government.

Tokyo Metropolitan Government (2013), *The Tokyo Cap-and-Trade Program Achieved 23% Reduction in the Second Year*, Tokyo: Tokyo Metropolitan Government.

UNFCCC (2010), 'Note Verbale', accessed 6 November 2014 at http://unfccc.int/files/meetings/cop_15/copenhagen_accord/application/pdf/japancphaccord_app1.pdf.

van Asselt, H.N. Kanie and M. Iguchi (2009), 'Japan's position in international climate policy: navigating between Kyoto and APP', *International Environmental Agreements: Politics, Law and Economics*, **9**(3), 319–36.

Vogel, D. (2003), 'The hare and the tortoise revisited: the new politics of consumer and environmental regulation in Europe', *British Journal of Political Science*, **33**(4), 557–80.

Watanabe, R. (2011), *Climate Policy Changes in Germany and Japan: A Path to Paradigmatic Policy Change*, London: Routledge.

World Bank (2014a), 'Japan', accessed 6 November 2014 at http://data.worldbank.org/country/japan.

World Bank (2014b), 'GINI index', accessed 6 November 2014 at http://data.worldbank.org/indicator/SI.POV.GINI?page=1&order=wbapi_data_value_2009%20wbapi_data_value%20wbapi_data_value-last&sort=asc.

World Bank (2015), 'Access to electricity (% of population)', accessed 6 May 2015 http://data.worldbank.org/indicator/EG.ELC.ACCS.ZS.

7. Russia's climate policy

Anna Korppoo*

BOX 7.1 KEY FACTS ABOUT RUSSIA

Population 2013	143.5 million
Population growth rate	Average 0.3% per annum 2010–13
Population density	9 per km^2
GDP, 2013	2096.777 USD billion
GDP/capita, 2013	14 612 USD
GDP growth rate 2013	1.3%
GINI index 2009	39.7
Access to electricity 2010	100% of the population
Land area	16 376 870 km^2
Political system	Federal, officially democracy but in practice autocracy
Subnational entities	Since 18 March 2014, the Russian Federation consists of 85 federal subjects, although the two most recently added subjects (Sevastopol and Crimea) are internationally recognized as part of Ukraine

Sources: World Bank (2014a–k); Russian Presidential Executive Office (2009).

INTRODUCTION

As a result of the social stresses of the post-Soviet period the population fell from 148.6 million in 1991 to 141.9 in 2009, slowly climbing to 143.5 million by 2013 (World Bank, 2014a). Officially, decision-making power is divided between the Duma and the Chief Executive (the president) under a democratic system; however, in practice President Putin rules under an autocratic regime (Box 7.1).

Russia is a high-income country (GDP per capita USD 14 612 in 2013) and was the world's eighth-largest economy in 2013 (World Bank, 2014h). According to Credit Suisse (2014), 85 per cent of this wealth is in the hands of the top 10 per cent of the population; however, based on the Gini index, the World Bank (and many others) consider wealth distribution in Russia to be basically similar to that of the United States or the United Kingdom (Gini index 39.7, 41.4 and 38.0, respectively) (World Bank, 2014f) (Box 7.1).

Russia's GDP grew by 3 to 4 per cent per annum between 2010 and 2012, dramatically slowing down in 2013 to 1.3 per cent (World Bank, 2014e). In May 2014, the Russian government approved an estimate of 0.5 per cent GDP growth for 2014 and 2 per cent in 2015 (BOFIT, 2014; *BOFIT Weekly*, 2014a). Repercussions of the conflict with Ukraine have been obstructing economic growth since summer 2014. At the time of writing (December 2014), Ural crude oil was trading at below USD 60 per barrel – down from USD 100 to 110 during the first half of 2014 – which has added to the slowdown (Neste Oil, 2014). BOFIT (*BOFIT Weekly*, 2014b) estimates that even though the Russian budget has been calculated on the assumption of an oil price at USD 100/barrel, the economy will be seriously affected only if the price should remain at USD 80–90 for over two years. A lower oil price leads to less economic activity, which also serves to slow down the growth in GHG emissions.

Russia's energy mix consists primarily of fossil fuels: in 2012, natural gas accounted for 51 per cent of total primary energy supply, oil 22 per cent and coal 18 per cent. The main non-fossil fuel source is nuclear power (6 per cent), with hydropower accounting for 2 per cent and waste and other renewable sources 1 per cent (IEA, 2012) (Table 7.1). As the country's fossil energy resources are domestic, and there has been no serious concern about the impacts of climate change on Russian territory, the only reason to limit their use is to secure supplies for export. This is likely to remain so for many years to come, as renewable energy (beyond large Soviet-built hydro), however significant the potential might be, remains in its infancy in Russia, and any significant increase in nuclear power seems unlikely in the short- or mid-term. *Energy Strategy of Russia for the Period up to 2030*, a document with a forecasting rather than planning character, even foresees the use of coal by 2030, to replace some of the currently dominant gas (Ministry of Energy of the Russian Federation, 2010).

Russia's GHG emissions were 2295 $MtCO_2e$ in 2012, fourth highest in the world, while the country's per capita CO_2 emissions were 12.2 tCO_2e in 2010 (World Bank, 2014g) (Table 7.1). Russia's current GHG emissions originate chiefly from electricity and heat generation, fuel industries, transport and industrial processes. Between 2008 and 2011, the transport sector (accounting for roughly 10 per cent of total emissions) was responsible for three-quarters of the growth in total emissions; CH_4

emissions from the oil and gas sector accounted for the rest of the increase (UNFCCC, 2014b).

Under the Kyoto Protocol, Russia has committed to limiting its emissions below the 1990 level. The emissions decline resulting from economic restructuring and closure of unprofitable economic activities in the aftermath of the Soviet Union left Russia's emissions significantly below this target level. Slow growth in emissions resumed from the late 1990s; between 2000 and 2012, average annual growth was 0.95 per cent (UNFCCC, 2014a). Russia rejected the second commitment period of the Kyoto Protocol, but in September 2013 passed a domestic GHG emissions limitation target aimed at keeping emissions growth 25 per cent below 1990 level until 2020. This target will require little additional policy action, as it actually allows for significant growth in emissions; in 2012, Russia's emissions were 32 per cent below the 1990 level (UNFCCC, 2014c).

Existing policies and measures with the potential to deliver emission reductions as side benefits include limitations on flaring associated petroleum gas (APG), and policies on energy efficiency and renewable energy. Implementation of the latter two has been slow, whereas more significant results have been achieved in limiting APG flaring. Regardless, Russia's main emitting sectors – energy and industry – and the main emission growth sector – transport – remain basically unregulated as regards CO_2 and other greenhouse gases. Thus, the ambition level of Russia's climate policy must be considered as low. Economic recession is more likely reason for a temporarily stabilizing emission trend in the future. Recent Russian emission modelling exercises have shown that downscaling GDP growth expectation from some 3–5 per cent down to some 2 per cent reduces the expected growth in emissions from some 80 per cent of 1990 level by 2020 to 70–75 per cent (Korppoo and Kokorin, forthcoming, 2015).

There is no particular domestic conflict concerning the goals of climate policy; most stakeholders agree that the emission decline in the 1990s should be seen as Russia's fair share of the global mitigation effort. It is also widely felt that other major emitters will have to adopt legally binding targets in order for Russia to do so, and that Russia should gain from participation in international climate regime. Russia's image as a foreign policy actor as well as the benefits available from participation in the international climate regime have been the main factors driving decision-making related to climate policy (Korppoo et al., 2015). 'Window dressing' policies appear to have been adopted in order to impress foreign partners: such policies have often remained unimplemented for a long time (Kokorin and Korppoo, 2013).

Table 7.1 Emissions and energy data for Russia

Total GHG emissions	1990	2010	2012
	3367.8 $MtCO_2e$[a]	2223.5 $MtCO_2e$	2297 $MtCO_2e$
GHG emissions per capita 2012	16.0 tCO_2e[b]		
GHG emissions change rate 2008–12	0.85%		
GHG emissions by sector 2012	Energy industries 39.9%, fugitive emissions 17.9%, transport 10.5%, industrial processes 7.9%, manufacturing industries and construction 6.9%, agriculture 6.3%, residential 4.6%, waste, 3.7%, other 2.3%		
Energy mix 2012	Coal 17.6%, oil 22.3%, gas 51.1%, nuclear 6.2%, hydro 1.9%, renewables and waste 1%		

Notes:

a. Million metric tons CO_2 equivalents.
b. Tons CO_2 equivalents.

Sources: IEA (2012); UNFCCC (2014a); World Bank indicators – population, total.

THE DEPENDENT VARIABLE: CLIMATE CHANGE MITIGATION POLICY

The Climate Doctrine was adopted in 2009 (President of the Russian Federation, 2009). It is politically and socially significant, as it officially recognizes the problem of climate change as well as its expected negative impacts on Russian territory. However, neither the document nor its subordinate Plan of Action offers specific new goals or policies and measures for adaptation or mitigation. Instead (and retrospectively listed), the actual policies and measures have been designed primarily to address other goals, like the conservation of energy resources and the indigenous development of clean technologies.

In the lead up to COP 19 in Warsaw in 2013, a voluntary economy-wide target to keep emissions 25 per cent below the 1990 level was adopted. As Russia's emissions in 2012 had remained 32 per cent below this level, the proposed target lacks environmental ambition, and represents more or less a 'business-as-usual' situation in emissions (see Bashmakov, 2009; McKinsey & Co., 2009; Malakhov, 2010, pp. 298–302; Ministry of Economic Development, 2013, p. 147).

The Federal Law on Energy Efficiency and Energy Savings, No. 261-FZ, was adopted in November 2009. Among its goals is to put into practice Edict No. 889 of 2008 (President of the Russian Federation, 2008), which established the target of cutting the energy intensity of the Russian economy by 40 per cent between 2007 and 2020. The law covers energy labelling, energy metering, energy audits, a ban on most incandescent light bulbs, measures on building material efficiency, energy-saving contracts and tariff and fiscal incentives. However, implementation of energy efficiency legislation has experienced problems, and it seems increasingly unlikely that the target will be achieved. For the period 2007 to 2011, the total primary energy supply (TPES) per GDP ratio actually increased by some 2.5 per cent (IEA/OECD, 2013), indicating that the mid-term decarbonizing effect of this law will be low.

Associated petroleum gas (APG), a side product of oil extraction, has traditionally been flared as waste. However, it can be utilized locally as fuel and raw material for the chemical industry, re-injected to increase pressure in the oil field, or collected and transported by pipeline to other users, reducing GHG emissions as a side benefit. In 2009, Government Decree 7 (Government of the Russian Federation, 2009a) introduced a 5 per cent limit to flaring of APG from 2012, with fines for exceeding this limit. In 2012, Government Decree 1148 (Government of the Russian Federation, 2012) raised the fines but also introduced significant exemptions to the rules, like exempting oil fields with small emissions, flaring during maintenance and new oil fields. Rostekhexpertiza (Aksenov et al., 2013) estimates that as a result some 60 per cent of APG can be flared without fines.[1] While some companies, especially privately owned ones, reported compliance with the 5 per cent rule in 2012, and others were improving their performance, state-owned companies (Rosneft, Gazprom Neft) blatantly violated the law by flaring some 30–50 per cent of the APG they produced (Kiryushin et al., 2013, pp. 15, 27). This goal, even though watered down, is a serious driver of change in the practices of the oil industry, but is less significant on the country level as regards GHG emissions, since emissions from APG represent only 1.8 per cent of the total (2012) (UNFCCC, 2014b).

In comparison to many countries with less significant sources of renewable energy (RES) available, Russia's energy mix is heavily dominated by fossil fuels, with only 1 per cent of TPES coming from 'new'

RES and waste, and a further 1.9 per cent from large Soviet-built hydro. According to the 2009 Governmental Order on Renewable Energy (Government of the Russian Federation, 2009b), Russia aimed for 11 GW of new renewable generation capacity and expected to generate 4.5 per cent of its electricity from renewables by 2020. A new law in May 2013 (Government of the Russian Federation, 2013) introduced solar, wind and small-scale hydropower generation to the Russian capacity market[2] in order to guarantee a 12–14 per cent return on investment over 15 years. The first tender was held in September 2013 and the second in July 2014; both were dominated by solar power, in part because of the domestic content requirements that are difficult for small hydro or wind power to fulfil (Roca, 2013; Rice et al., 2014). As less than 1 per cent of electricity was generated from renewable sources of energy in 2011, the stepwise target of 1.5 per cent by 2010 was clearly not met, and the target of 2.5 per cent by 2015 was in danger (Jegelevicius, 2013) when the government downgraded the target to 2.5 per cent by 2020 (Leschinskaya, 2013). We must conclude that RES policy does not feature on the main agenda of the Russian government, and cannot be expected to make an observable difference in decarbonizing the economy in a mid-term perspective.

Some further climate-focused policies have been given the go-ahead from the top political level in 2014. It is foreseen that the domestic offsets scheme will use the system established for administering Kyoto Joint Implementation (JI) projects to serve as its basis. Details such as the state acting as the main buyer, as well as measures to ensure that the scheme will successfully direct the funds to real projects, are under discussion. The preparation of a domestic ETS has also been approved. However, given its political nature as regards domestic burden sharing as well as the sole focus on climate mitigation policy, the mechanism might be able to deliver emission reductions only after 2020.

It is more important to see what is missing when evaluating the ambition level of Russia's policies for dealing with climate change. Its energy efficiency policy focuses only on less significant energy users – households and the public sector – and not the main users, which are energy and industry. The transport sector, accounting for 95 per cent of Russia's GHG emission growth, is not covered by any climate policy. Further, the energy mix is foreseen to remain based on fossil fuels; in order to secure sufficient gas resources for export, the leadership is even promoting coal to replace gas in domestic consumption (Ministry of Energy of the Russian Federation, 2010, p. 40).

The difference in approaches to climate change has caused problems for climate cooperation between Russia and many OECD members, especially the EU. The latter sees the issue as urgent and real, whereas Russia has remained sceptical to both its urgency and its significance. This has led to Moscow mimicking Western approaches for reasons other than climate protection, and not engaging in mitigation actions driven by genuine environmental concerns. From the Russian perspective, the country has achieved its share of emission reductions for now, thanks to the post-Soviet emission decline, and can even be considered a frontrunner. As a result, Moscow regards it as fair to wait with further mitigation actions until other countries act. This contrasts with the view widely held abroad – based on the virtual absence of Russian mitigation policies and the country's poor track record of energy efficiency – that Russia has done little to decarbonize its economy and is thus a laggard in climate mitigation efforts. Finally, Russia heads the Eastern European group in UNFCCC climate negotiations; the group has profiled itself as pushing for market mechanisms, especially in view of the post-Soviet surpluses also beyond the Annex I group of countries.

EXPLAINING POLICY TRAJECTORIES

Fundamental Parameters

Material parameter

According to the *Energy Strategy of Russia for the Period up to 2030*, Russia holds some 23 per cent of world gas reserves and 19 per cent of coal reserves (Ministry of Energy of the Russian Federation, 2010, p. 21). Its oil reserves were estimated to account for some 5 per cent of the world reserves in 2013 (EIA, 2014). Russia's renewable energy resources are vast; the IEA has estimated the economic potential as equivalent to some 25 per cent of the country's 2012 total primary energy supply (OECD/IEA, 2003).[3]

Domestic energy resources form one of the main interest clusters of the political leadership, as 17–25 per cent of Russia's GDP has been estimated to originate from the oil and gas sectors including exports, with the total energy sector accounting for some 30 per cent of GDP (Simola, 2013). Ample gas supplies and the policy of subsidizing households as well as industrial consumers of gas and electricity (generated mainly at gas-fuelled power plants) have kept domestic gas prices relatively low and contributed to a 'systemic lock-in' to a carbon-based fuel economy. But also the infrastructure – still largely reflecting the principles of a centrally planned economy – has limited the scope for change in energy mix. Examples here include the limited possibilities for switching between different fuels, and lack of gas/electricity metering in many places. There has been some (economically based) interest in cutting domestic energy waste, but the monopolistic structures of the economy have not been conducive to reform of consumption, and certainly having relatively cheap gas readily available has led to glaring waste. Gas prices, which are controlled, have been raised considerably over the last ten years, primarily to secure funds for investment in new production. Price increases were frozen as a result of economic recession in 2012 (Leifheit, 2014). The rigidity of consumption can be seen in the fairly limited effect of price hikes on demand. Serious change will also require major investments in technologies for consumption. The gas industry remains underfinanced, with no comprehensive reform of consumption in sight.

The nuclear sector has political support, including plans to supply up to 20 per cent of electricity by nuclear generation by 2030, in comparison to 16 per cent in 2008 (Ministry of Energy of Russian Federation, 2010, p. 148). Since the 1990s low point in investments, several new reactors have been commissioned, and Russian nuclear technology is being exported. However, it is difficult to judge how significant the role of nuclear power could become in the mid-term, as the Soviet-built nuclear capacity is nearing the end of its lifetime and will have to be decommissioned.

There are also plans for expanding large hydropower in Russia. However, they may have been influenced by the 2009 disaster at the Sayano–Shushenskaya hydroelectric station, in which 75 people died (Hasler, 2010).

Obsolete Soviet-inherited infrastructure, on the supply as well as the demand side, continues to create a 'technological lock-in' to a carbon-intensive system even though much of this capacity has been closed down. In many cases, Soviet-era technologies and heavy industry have now exceeded their planned technical lifetime. This adds to the energy intensity of the economy and may even cause industrial disasters, as with the Sayano–Shushenskaya explosion. In many locations, the social importance of some facilities – a 'social lock-in' to a carbon-intensive economy – is an important factor. Moreover, obsolete infrastructure remains in use because of the uncertain investment environment, which blocks many modernization investments for upgrading production facilities.

Some 60 per cent of the territory of the vast Russian Federation is affected by permafrost, which makes all infrastructure vulnerable to melting as a result of climate change; pipelines, roads and the northern settlements are in particular danger (Anisimov and Reneva, 2006). In the Soviet Union, public transport was the main mode of mobility, and has remained significant in urban areas. However, private cars have taken over a major share of transport in cities, and the same trend is now seen also in towns and rural areas where public transport tends to be poor. Moreover, the size of private cars is increasing. These emission growth trends are expected to continue at least until 2020 (Yulkin, 2014). Most oil and gas is transported in pipelines. This is a carbon-efficient mode of transportation as such, but leaks in the Russian pipeline network add to the atmospheric burden.

As internationally agreed limits to GHG emissions may reduce demand for Russia's oil and gas exports over time, one would expect the interest in the continuity of energy exports to undermine mitigation policies and participation in international climate regimes (note, for instance, Saudi Arabia). However, as policies that can deliver emission reductions have remained inefficient and do not cover the major emitters – the energy and industry sectors – the link between fossil fuel industries and domestic action is weak in Russia. Further, the short-term approach to policy-making, combined with scepticism as to the likelihood that a new and effective international regime will be established, reduce worries over the impacts of climate policies on Russia's fossil-fuel exports.[4] Finally, the domestic justifications to business-as-usual emissions commitments add to the attractiveness of international climate cooperation as an arena for soft foreign policy.

Political parameters

Russia's political system is built around a strong president, elected for a six-year term. The president appoints the prime minister (with the consent of the Duma), approves the members of the government proposed by the prime minister, and directs the activities of the government. The president also signs all acts. According to the Constitution, most legislative power belongs to the Federal Assembly. In reality, however, decision-making has become concentrated in the hands of President Vladimir Putin and his party of power, United Russia, even if constitutional procedures are generally followed. This has been called 'a theatre democracy' as the organs of democratic power have in practice been subsumed under the president, whether through democratic or undemocratic means. The former concerns especially coalition building in the Duma, whereas the latter includes changes in the structures within the Duma, excluding candidates from elections, suffocating free media and distorting election results.

The regions have an important role in policy implementation but are virtually absent from climate-related decision-making, which takes place at the federal level. The low and uncertain financial allocations earmarked to regional level mitigation actions by the federal government constitute a major barrier..

Ministries and agencies typically lack the Weberian objectivity of the public service. Instead of producing public goods, they often generate private and network profits, also known as 'administrative capture' (Ledeneva, 2013). In some cases networks pick initiatives that can push climate change–related action to the agenda. JI provided an example of such a foreseen rent, with the interest from the administration that followed.

The debate among policy-makers and the public alike typically balances between the positive and negative impacts of climate change on Russian territory. For instance, the Arctic territories are warming faster than the global average, placing river banks and settlements built on permafrost in danger – but the melting sea ice brings new economic opportunities such as the use of the Northern Sea Route and utilization of Arctic fossil fuel deposits. The contribution of natural climatic changes is strongly present in the debate: not all impacts of climate change are always seen as induced by human activity. These starting points reduce the political impact of expected climate change in the natural and urban environments. Even if some dramatic developments are forecast – such as increases in forest fires, melting permafrost, sea-level rise and flooding – they are unlikely to trigger mitigation action.

Driving Forces

Governmental supply

The Russian presidential administration and the government appointed by the president are the main sources of policy supply in terms of decision-making, while the ministries and agencies under particular ministers are key actors in terms of policy design, approval and implementation. The main ministries and agencies relevant to climate policies are the Ministry of Economic Development, Roshydromet, the Ministry of Natural Resources and the Ministry of Energy, as well as the Ministry of Foreign Affairs in formulating Russia's international climate position.

Expert organizations, business and NGOs cannot generate demand for policies as can their colleagues in democratic systems. However, they contribute to policy supply by feeding ideas to the official preparatory processes, and by building coalitions with like-minded actors within the bureaucracy.

Policy initiatives can be divided into those focused solely on climate goals, and those that deliver emissions reductions more as a side effect. Climate-focused goals have not yet led to additional implementation; the

Climate Doctrine did not introduce any new mitigation measures and the 2020 goal remains just a number. Policies with the potential to deliver emissions reductions on the side (energy efficiency, renewable energy and APG) and JI follow basically the same pattern. They tend to consist of a declared goal, long – sometimes stalled – preparatory process, serious problems with implementation and watered-down results behind schedule.

What have been the main drivers of climate policies in Russia? *Foreign policy interests* centre on Russia's image as a serious negotiation partner. Moscow's desire to be accounted to the club of major economies has led to passing 'window-dressing' policies, to have something to show in the international arena. Issue linkages have also driven climate policy: gaining EU support for Russian WTO membership lent support to the Kyoto Protocol in Moscow. The *economic interests of domestic actors* in climate policies, mostly in the Flexible Mechanisms of the Kyoto Protocol, have led to the adoption of some mitigation-related measures. However, such coalitions are difficult to bring together in the Russian political culture because the potential economic benefits tend to be minor compared to the profits of oil, gas and big industries. Finally, *policy goals unrelated to climate change* – for instance, economizing energy use and modernization of the economy – have often led to the launch of policies with the potential to deliver emissions reductions as a side effect. Indeed, this might prove the most important driver, if the administrative barriers could be solved.

And what then are the main barriers to decarbonizing the Russian economy? The *abundance of domestic fossil fuel resources* reduces pressures related to energy security, and thus, the urgency in developing renewable energy sources. Further, lock-in to the existing fossil fuel infrastructure, as well as the economic importance of fossil fuels as an export sector, make it unlikely that any massive decarbonization plans could gain political support in Russia. The *lack of political will* has been evident in the low ambition of national-level mitigation targets, and formulating the country's climate policy position on the basis of the 1990s emissions decline. Further, domestic policies appear to have a token character as regards mitigation: the main emitting and emission growth sectors are not covered, and the goals set are close to a business-as-usual trend, except for the APG 5 per cent rule. *Weak implementation* of launched policies is endemic in the Russian political culture. In the case of low-priority issue areas, stated goals as well as actual designed policy measures included in the legislation may remain unimplemented for years, unless an advocate strong enough to elevate the policy to the political agenda emerges. Policy design tends to be weak, because of the heavy bureaucracy surrounding the preparatory process. This allows various interested parties to get elements of their choice included, so as to gain bureaucratic and private benefits. Further, state companies can allow themselves some leeway in reaching the goals set.

Societal demand

Policy cycle dynamics beyond Putin's dominance remain insignificant. Most real demand for policies is expressed in non-democratic processes behind closed doors. 'Official' demands by the Duma and the public discussion are either muted or ignored by the power elite. In Tsebelis's terms, the Duma has been absorbed as a 'veto player' (Tsebelis, 2002) by the president, becoming more or less a rubber stamp used for legitimizing his autocratic rule, and that of his elite.

Willingness to pay for climate mitigation – as the main goal of a policy – is almost absent. Thus, there has been little societal demand for costly mitigation policies, whether latent or manifest. Insufficient information on the regional impacts of climate change and the related costs is used to justify blocking climate-based initiatives in the Russian political system, since cost–benefit comparisons are seen as the starting point for establishing a policy, at least on the rhetorical level. That means that 'demand' for policies that can deliver emission reductions can thus only be unrelated to concern about the impacts of climate change.

Even if the political system were more receptive to democratic demands, there would not be much pressure to halt climate change. The issue is off the agenda of the average Russian, and public pressures on the country's leadership focus more on improving the quality of life. Among environmental issues, priority goes to local problems, especially those influencing the health of the population. According to Russia's leading polling organization, the Levada Center, about half of those surveyed in its annual polls 2011–14 ranked the pollution of water and air as the most serious environmental problems, while climate change was deemed most serious by only some 21–24 per cent (Levada Center, 2014).

A company's ability to lobby decision-makers depends on its status. State industry and members of the president's inner circle involved in industrial activities are far better positioned to influence policies than private industry, and larger industrial actors tend to have more leverage than smaller entrepreneurs. When the economic interests of the president and the industry coincide, there is more space for negotiation. As regards climate mitigation, however, the more influential industries seldom push for low-carbon policies.

Some demand for specific policy solutions is generated by the executive level in public administration (ministries, agencies) if it can further the interests of actor networks ruling individual agencies. Although no permanent climate coalition exists in Russia, policy-specific coalitions of interested individuals have promoted particular policy options, with some success. For instance, key individuals moved into relevant administrative positions in the Ministry of Economic Development and Sberbank to push for establishing a functional domestic JI administration (Korppoo et al., 2015).

Economic growth is the main driving factor behind all policy-making on the part of the Russian leadership. Aimed at keeping the public satisfied, such economic growth is secured through energy export revenues. The success of this policy has tended to lead to growth in GHG emissions – but, with the current economic turmoil, the years to come may see a flat emissions trend.

Technological advance is a key driver of policy – or at least rhetoric – for the Russian leadership; it has been flagged by the modernization policy launched by President Medvedev. The main motivation is to keep pace with other geopolitically important players while updating technologies: this also makes economic sense, as it helps to reduce the prevailing waste of energy and resources. Modernization policy can support actions that pave the way towards a lower-carbon economy. Indeed, the promoters of mitigation policies have often emphasized such linkages.

Supply–demand relationship

Any significant positive policy change seems unlikely in Russia, as the current supply of mitigation policies by the government can be argued to meet societal demand. Governmental supply and societal demand are insignificant in terms of policies driven by concerns for climate warming. Some supply and demand can be identified for policies that can deliver emission reductions as a side effect, both in the legislative and the implementation phases. The relevance of this can be questioned, however, since statute level policy is likely to become watered down during implementation.

POLICY TRAJECTORY

The prospects for significant policy change towards decarbonization are low in Russia. To a considerable extent, the low demand for mitigation policies can be explained by the lack of environmental concern, and the interpretation of the surplus originating from the post-Soviet emissions decline recognized as a climate effort.

In the Russian debate, the expected impacts of climate change are largely detached from mitigation actions. Changing this would require firmly establishing stronger links between human action and negative changes in climatic conditions. The Russian approach to science seems unlikely to follow the West in its 'consensus' on human activities as the main cause of climate change, with the impacts seen as generally negative, and policies intended to remedy these negative effects. Thus, it seems highly unlikely that a demand for costly domestic mitigation measures will emerge in Russia. Even if that should occur, any wider decarbonizing efforts would rapidly come into conflict with the goals of the country's economically vital fossil fuel sector.

The main source of change can be expected to emerge as a result of political goals other than mitigation, for instance modernization and cutting energy waste, as well as private and network interests of powerful individuals and lobbies. However, many factors act to obstruct such policies. The domestic supply of fossil fuels makes it seem less urgent to implement resource savings, in comparison to countries with drivers related to security of supply. This is particularly true for gas, where Russian supplies have been sufficient to cover both domestic and export demand. Oil usually has an alternative export outlet to domestic consumption, but is nevertheless affected by the same logic since the large domestic production takes pressure (psychological and political) off energy conservation. The investment climate is difficult for businesses and households, which reduces the likelihood of dramatic updates of the country's obsolete infrastructure in the short and medium term. The economic havoc unfolding at the time of writing can only make the situation worse, at least in a medium-term perspective. Finally, the problems of policy implementation affecting the Russian political system are so grave that, even when adopted, policies that do not happen to be at the top of political agenda tend to get stuck or watered down.

Until 2020, Russia's GHG emission trend is likely to remain more or less flat. The domestic target of keeping GHG emissions to 75 per cent of the 1990 level until 2020 was not much beyond business-as-usual when it was passed. In view of the lowered expectations of GDP growth following the plummeting oil price, achieving the 75 per cent target is likely to be even easier, despite the failure of existing domestic policies. But in the longer term, GDP growth will resume, and GHG emissions will follow suit. No focused policies have been introduced to target the main sectors of emissions growth: transport and gas transport. Their growth trends are likely to continue.

Price increases in the dominant fuel, gas, could trigger wider saving of energy waste and bring increased pressure for modernizing the technological basis by gradually pushing investments over the cost-efficiency threshold. In the short term, however, the economic downturn is likely to continue stalling regulated price increases.

Changes in Russia's geopolitical role have come as a result of the conflict with Ukraine, with various political and economic sanctions introduced against Russia by the EU and the USA. An extreme scenario could be a popular – or a palace – overthrow of the Putin regime as a response to the dwindling economic growth resulting from the sanctions that affect the future prospects of the population or the economic elite. At the time of writing, Putin has retained strong popular support regardless of the economic and other problems that the Russian people are experiencing. Given the lack of environmental concern about the impacts of climate change and popular support to costly mitigation policies, even radical political changes are unlikely to push for progressive decarbonizing policies in Russia.

On the international level, Russia feels that it has already done its fair share as its GHG emissions plummeted in the aftermath of the Soviet economic restructuring in the 1990s, and have remained well below 1990 level even today. Alongside its conflict with Ukraine, Putin's regime has been taking a new, more aggressive course in the international political arena. Moscow is likely to wait and see what will happen with other main players as to the future international climate regime.

NOTES

* This work was funded by the post-doctoral research project 252853 of the Academy of Finland and the NORKLIMA project 207810 of the Research Council of Norway. Special thanks go to Arild Moe for his contribution to the chapter.

1. The coverage of Decree 1148 excludes some 18–19 per cent of APG flaring (including the 5 per cent that is allowed) due to measurement errors, maintenance and low quality of APG. Further, exemptions for 'small' and 'new' oil fields account for 30–40 per cent of the total.
2. Capacity market ensures security of electricity supply by providing a payment for reliable sources of capacity, alongside their electricity revenues, to ensure they deliver energy when needed. It also generates a longer-term price signal to support capacity investments by electricity producers.
3. Author's calculations: the estimated economic potential 273.5 $MtCO_2e$ per annum (OECD/IEA, 2003, p. 30), Russia's TPES 2012 1080.8 $MtCO_2e$ (IEA Energy Balances).
4. This is not the case with the appearance of new sources of fuels, which may take place far more quickly; considerable concern has been expressed about the shale gas 'revolution', and its impacts on Russia's energy exports. See *The Economist* (2013).

REFERENCES

Aksenov, A., V. Skobelina and I. Tremasova (2013), *Dolgo 'zapryagali'– poedem li?* [Slow to Harness, Are we Going?], undated report available (in Russian) at the Rostekhexpertiza website, accessed 25 September 2013 http://www.rostehexpertiza.ru/analytics/dolgo-zapryagali-poedem-li/.

Anisimov, O. and S. Reneva (2006), 'Permafrost and changing climate: the Russian perspective', *AMBIO*, **35**(4), 169–76.

Bashmakov, I. (2008), *Nizkouglerodnaya Rossiya: 2050 god* [Low-carbon Russia: 2050], Moscow: Center for Energy Efficiency (in Russian).

BOFIT (2014), 'BOFIT Russia statistics', accessed 3 June 2014 at http://www.suomenpankki.fi/bofit_en/seuranta/venajatilastot/Pages/default.aspx.

BOFIT Weekly (2014a), 2014/22, 30 May 2014, 'Russia', accessed 3 June 2014 at http://www.suomenpankki.fi/bofit_en/seuranta/viikkokatsaus/Documents/w201422.pdf.

BOFIT Weekly (2014b), 2014/43, 24 October 2014, 'Lower crude oil prices not yet fatal for Russia's fiscal or current account balances', accessed 25 November 2014 at http://www.suomenpankki.fi/bofit_en/seuranta/seuranta-aineisto/Pages/vw201443_3.aspx.

Credit Suisse (2014), Table 4-4 in *Global Wealth Databook 2014*, accessed 24 November 2014 at https://publications.credit-suisse.com/tasks/render/file/?fileID=5521F296-D460-2B88-081889DB12817E02.

Government of the Russian Federation (2009a), Decree No. 7, 8 January 2009, 'On measures to stimulate the reduction of air pollution from associated gas flaring products' (in Russian).

Government of the Russian Federation (2009b), Edict No. 1-r dated 8 January 2009, 'About government policy on energy efficiency on the basis of renewable energy sources until 2020' (in Russian).

Government of the Russian Federation (2012), 'Decree No. 1148', dated 8 November 2012, 'On the particularities of calculating fees for the emissions of polluting substances that are formed during burning at flare facilities and (or) of dispersion of associated gas' (in Russian).

Government of the Russian Federation (2013), Decree No. 449 dated 28 May 2013, 'On the mechanism for stimulating use of renewable energy on the wholesale electric power and capacity market' (in Russian).

Hasler, J. (2010), 'Investigating Russia's biggest dam explosion: what went wrong', *Popular Mechanics*, 2 February, accessed 14 December 2014 at http://www.popularmechanics.com/technology/engineering/gonzo/4344681.

International Energy Agency (IEA) (2012), 'Russian Federation: balances for 2012', accessed 25 November 2014 at http://www.iea.org/statistics/statisticssearch/report/?country=RUSSIA&product=Balances&year=2012.

International Energy Agency/Organisation for Economic Co-operation and Development (IEA/OECD) (2013), 'IEA beyond 2020 energy balances for non-OECD countries', accessed 26 November 2014 at http://wds.iea.org/WDS/Common/Login/login.aspx (subscription required).

Jegelevicius, L. (2013), 'Russian renewable energy law could significantly boost clean energy but obstacles remain', *RenewableEnergyWorld.com*,

8 May, accessed 26 November 2014 at http://www.renewableenergyworld.com/rea/news/article/2013/05/new-russian-renewable-energy-source-development-measures-to-overcome-obstacles.

Kiryushin, P., A. Knizhnikov, K. Kochi, T. Puzanova and S. Uvaro (2013), *Poputny neftyanoy gaz v Rossii: 'Szhigat'nel'zya, pererabatyvat'!' Analiticheskij doklad ob ekonomicheskikh i ekologicheskikh izderzhkakh czhiganiya poputnogo neftyanogo gaza v Rossii* [Associated Gas in Russia: 'Don't Flare, Utilize!' Analytical presentation of economic and environmental costs of associate gas flaring in Russia], report by WWF, Moscow.

Korppoo, A. and A. Kokorin (2013), *Russia's Post-Kyoto Climate Policy: Real Action or Merely Window-Dressing?*, FNI Climate Policy Perspectives 10 (May), Lysaker, Norway: Fridtjof Nansen Institute.

Korppoo, A. and A. Kokorin (2015), 'Russia's 2020 greenhouse gas emissions target: success factors', *Climate Policy* (forthcoming).

Korppoo, A., N. Tynkkynen and G. Hønneland (2015), *Russia and the Politics of International Environmental Regimes. Environmental Encounters or Foreign Policy?*, Cheltenham, UK and Northampton, MA, USA: Edward Elgar Publishing.

Ledeneva, A. (2013), *Can Russia Modernize? Sistema, Power Networks and Informal Governance*, Cambridge, UK: Cambridge University Press.

Leifheit, D. (2014), 'Russian gas: domestic market blues', *Natural Gas Europe*, 14 July, accessed 3 January 2015 at http://www.naturalgaseurope.com/russian-gas-market-tatiana-mitrova.

Leschinskaya, K. (2013), 'Renewable year-end: focus Russia', *RenewableEnergyWorld.com*, 31 December, accessed 31 December 2014 at http://www.renewableenergyworld.com/rea/news/article/2013/12/renewable-year-end-focus-russia.

Levada Center (2014), 'Ehologicheskie problemy i bezopasnost' [Ecological problems and security], 10 July (in Russian), accessed 1 September 2014 at http://www.levada.ru/10-07-2014/ekologicheskie-problemy-i-bezopasnost.

Malakhov, V. (2010), 'Economic perspectives on low-carbon development in Russia', *International Journal of Low-Carbon Development*, **5**(4), 298–302.

McKinsey & Company, Inc. (2009), *Pathways to an Energy and Carbon Efficient Russia*, accessed 30 September 2013 at http://www.mckinsey.com/~/media/mckinsey/dotcom/client_service/sustainability/cost%20curve%20pdfs/russian_cost_curve_summary_english.ashx.

Ministry of Economic Development (2013), *Prognoz dolgosrochnogo sotsial'nogo-ekonomicheskogo razvitiya Rossiyskoy Federatsii na period do 2030 goda* [Projection of long-term social and economic development of the Russian Federation until 2030], 25 March (in Russian), accessed 27 August 2013 at http://www.economy.gov.ru/minec/activity/sections/macro/prognoz/doc20130325_06.

Ministry of Energy of the Russian Federation (2010), *Energy Strategy of Russia for the Period up to 2030*, Moscow.

Neste Oil (2014), 'Crude oil prices', accessed 20 December 2014 at http://www.nesteoil.com/default.asp?path=1,41,538,2035,5193.

Organisation for Economic Co-operation and Development/International Energy Agency (OECD/IEA) (2003), *Renewables in Russia: From Opportunity to Reality*, Paris: OECD/IEA.

President of the Russian Federation (2008), Edict No. 889 dated 4 June 2008 'On Measures to make the Russian Economy more Energy and Environmentally Efficient' (in Russian).

President of the Russian Federation (2009), Edict dated 17 December 2009 No. 861-rp 'On the Climate Doctrine of the Russian Federation' (in Russian).

Rice, B., A. Blomfield and A. Rotar (2014), 'Russia's second renewable energy auction gives both glimmer of hope and cause for concern', *King & Spalding Energy News Letter*, July, accessed 26 December 2014 at http://www.kslaw.com/library/newsletters/EnergyNewsletter/2014/July/article1.html.

Roca, M. (2013), 'Russia awards first renewable energy tender to boost the industry', *RenewableEnergyWorld.com*, 26 September, accessed 3 February 2015 at http://www.renewableenergyworld.com/rea/news/article/2013/09/russia-awards-first-renewable-energy-tender-to-boost-the-industry.

Russian Presidential Executive Office (2009), *The Constitution of the Russian Federation*, accessed 9 May 2015 at http://constitution.kremlin.ru/#article-65.

Simola, H. (2013), 'Assessing the contribution of Russia's oil and gas sector to GDP', in H. Simola, L. Solanko and V. Korhonen, *Perspectives on Russia's Energy Sector*, BOFIT Online 3/2013, Helsinki: Bank of Finland Institute for Economies in Transition, accessed 3 May 2015 at http://www.suomenpankki.fi/pdf/172269.pdf.

The Economist (2013), 'Spooked by shale: the shale-gas revolution unnerves Russian state capitalism', *Economist.com*, 29 June, accessed 29 April 2015 at http://www.economist.com/news/business/21580131-shale-gas-revolution-unnerves-russian-state-capitalism-spooked-shale.

Tsebelis, G. (2002), *Veto Players: How Political Institutions Work*, Princeton, NJ: Princeton University Press.

UNFCCC (2014a), 'Summary of GHG emissions for Russian Federation', accessed 5 February 2015 at https://unfccc.int/files/ghg_emissions_data/application/pdf/rus_ghg_profile.pdf.

UNFCCC (2014b), 'Greenhouse gas inventory data – detailed data by party', accessed 5 February 2015 at http://unfccc.int/di/DetailedByParty.do.

UNFCCC (2014c) 'GHG data from UNFCCC', accessed 5 February 2015 at http://unfccc.int/ghg_data/ghg_data_unfccc/items/4146.php.

US Energy Information Administration (EIA) (2014), 'World energy statistics – crude oil proved reserves', accessed 26 November 2014 at http://www.eia.gov/cfapps/ipdbproject/iedindex3.cfm?tid=5&pid=57&aid=6&cid=regions&syid=2010&eyid=2014&unit=BB.

World Bank (2014a), 'Data: population, total', accessed 25 November 2014 at http://data.worldbank.org/indicator/SP.POP.TOTL.

World Bank (2014b), 'Data: population growth, annual %', accessed 25 November 2014 at http://data.worldbank.org/indicator/SP.POP.GROW.

World Bank (2014c), 'Data bank: GDP, 2013, current USD', accessed 25 November 2014 at http://data.worldbank.org/indicator/NY.GDP.MKTP.CD.

World Bank (2014d), 'Data: GDP per capita, current US$ [2013]', accessed 25 November 2014 at http://data.worldbank.org/indicator/NY.GDP.PCAP.CD.

World Bank (2014e), 'Data: GDP growth, annual %', accessed 26 November 2014 at http://data.worldbank.org/indicator/NY.GDP.MKTP.KD.ZG.

World Bank (2014f), 'Data: Gini index', accessed 17 November 2014 at http://databank.worldbank.org/data/indicator/SI.POV.GINI?id=af3ce82b&report_name=Popular_indicators&type=series&ispopular=y.

World Bank (2014g), '3.8 World Development Indicators: energy dependency, efficiency and carbon dioxide emissions', accessed 16 January 2015 at http://wdi.worldbank.org/table/3.8.

World Bank (2014h), 'World Development Indicators: GDP, current US$', accessed 22 November 2014 at http://data.worldbank.org/indicator/NY.GDP.MKTP.CD?display=default.

World Bank (2014i), 'Population density (people per sq. km of land area)', accessed 9 May 2015 at http://data.worldbank.org/indicator/EN.POP.DNST.

World Bank (2014j), 'Access to electricity (% of population)', accessed 9 May 2015 at http://data.worldbank.org/indicator/EG.ELC.ACCS.ZS/.

World Bank (2014k), 'Land area (sq. km)', accessed 9 May 2015 at http://data.worldbank.org/indicator/AG.LND.TOTL.K2.

Yulkin, M. (2014) (Russian Union of Industrialists and Entrepreneurs), *Sectoral Structure and Dynamic of GHG Emissions in the Russian Federation in 2011*, 21 January 2014, Round Table on Problems of GHG Regulation, Ministry of Economy.

8. The United States: Obama's push for climate policy change

Guri Bang*

BOX 8.1 KEY FACTS ABOUT THE UNITED STATES

Population 2013	316.1 million
Population growth rate 2013	0.7%
Population density 2013	35 per km^2
GDP per capita 2013	53 041 USD
GDP growth rate 2013	2%
GINI index 2010	41.1
Access to electricity	100% of the population
Land area	9 147 420 km^2
Political system	Federal republic
Subnational entities	50 states and the federal district

Source: World Bank data (2015a, 2015b).

INTRODUCTION

The position of the United States (Box 8.1) in international climate negotiations has for 25 years been one of caution towards binding international commitments to reduce GHG emissions for fear of negative domestic economic consequences. Although President Clinton signed the Kyoto Protocol in 1997, his successor George W. Bush rejected the treaty in 2001 with strong backing in the US Congress. Main arguments for withdrawing from Kyoto were the lack of participation by emerging economies that contribute increasing shares of world total emissions, and the negative consequences GHG emissions cuts under these conditions could have for the US economy. These two elements also made the prospects for an ambitious and economy-wide domestic climate policy programme bleak ever since the inception of the global warming debate in US politics.

Economic growth has traditionally depended on cheap, fossil fuel energy mainly provided from plentiful domestic reserves of coal. The USA is now the world's second-largest emitter of greenhouse gases, surpassed only by China (Table 8.1). Emissions have declined steadily over the past decade in step with the economic downturn, and the onset of the shale gas revolution. New horizontal drilling technologies opened huge reserves of shale oil and shale gas for exploration that further tied economic growth to the burning of fossil fuels. However, increased use of natural gas in the power sector resulted in declining levels of energy-related CO_2 emissions from 2007 onwards. The deep reliance on fossil fuel energy, in combination with intense polarization between Republicans and Democrats on climate policy issues, strong opposition to climate action from key stakeholder groups, and lack of public pressure to act on the problem are the most important barriers to an ambitious domestic climate policy in the United States.

Despite these profound barriers, the Obama administration proposed significant changes in the approach to climate change policy. In the first two years of his presidency, Obama pushed for the enactment of a federal climate law, but senators opposed to climate action blocked the initiative. After proposing a Climate Action Plan in 2013, the president employed his executive powers to develop new regulations aimed at cutting CO_2 emissions through existing law. As a result, state-specific CO_2 regulations for power plants are being developed for the first time in the United States under the jurisdiction of the Clean Air Act. Furthermore, Obama authorized stricter emissions standards for vehicles, and proposed policies to address the impacts of climate change. Finally Obama promised to take a leadership role in the international climate negotiations, striking a bilateral deal with China in 2014 whereby the two major emitters committed to reduce GHG emissions levels and phase in additional renewable energy in their respective domestic energy mixes.

This chapter explores the drivers and barriers for realizing the Obama administration's proposed climate policy changes. While public pressure for more federal climate action has slowly increased in step with the recovery of the US economy after the financial crisis, deep disagreements that have dominated US climate policy deliberations among federal law-makers over the past 20 years persist. The chapter concludes with an outline of likely policy trajectories up to 2020.

Table 8.1 *GHG emissions and energy mix in the United States*

Total GHG emissions by year	1990	2005	2012
	6233 $MtCO_2e$[a]	7253 $MtCO_2e$	6525 $MtCO_2e$
GHG emissions per capita 2012	17.6 tCO_2e[b]		
GHG emissions by sector 2012	Electricity 32%, transport 28%, industry 20%, agriculture 9%, commercial 5%, residential 5%		
Energy mix 2013 (primary energy consumption)	Natural gas 27%, coal 19%, petroleum 36%, renewable 10%, nuclear 8%		

Notes:

a. Million metric tons CO_2 equivalents.
b. Tons CO_2 equivalents.

Sources: Data for GHG emissions: US Environmental Protection Agency (2015). Data for energy mix: US Energy Information Administration (2015).

CURRENT CLIMATE POLICY AND RECENT TRENDS

Barriers to Federal Climate Policy in the Past

For many years the focus of attempts to change US climate policy was on legislative action, which met with fierce and constant opposition in the US Congress. Legislation proposing a cap-and-trade system was voted down repeatedly in the Senate in 2003, 2005 and 2008. When the Democrats took over the majority in both chambers of Congress in 2006, they initiated a campaign to put climate change on the congressional agenda. Hundreds of congressional hearings were held on the subject of federal action on climate policy, and many bills were proposed outlining various strategies for how federal policy could be designed cost-effectively (Bang and Schreurs, 2010). In his first year in office (2009), Obama elevated climate change to a top political agenda item. Climate legislation was introduced and debated in the 111th Congress (2009–10), and the House of Representatives passed the American Clean Energy and Security Act in June 2009 by a slim majority. However, the bill died in the Senate in 2010 without a floor debate (Wallach, 2012). No climate bills have been debated in Congress since then.

Two types of concerns that conflict with ambitious climate policy programmes have dominated energy and climate policy debates in the US Congress: concerns over energy security and sustained low energy prices have overshadowed concerns for environmental protection. Importantly, security for energy resource availability and low energy prices are the main priorities for big, energy-intensive industries and utilities (Bang, 2010). Energy security and low energy prices have traditionally been closely associated with coal resources and coal-fired electricity, because of abundant domestic coal reserves, and because coal prices have been low for decades (Fisher, 2004). Industry actors – organized in politically powerful lobbying groups such as the US Chamber of Commerce and the National Mining Association – have put political pressure on law-makers to secure a stable supply of low-priced energy, and opposition to strong climate policy programmes has moved in step with such demands.

Furthermore, over the past decade climate change has become an increasingly partisan issue in US politics. With the growth of Tea Party power within the Republican Party, congressional Republicans increasingly tended to vote against any proposed climate policy programmes (Skocpol, 2013). Democrats have consistently been more supportive, but have not been able to build a majority coalition in Congress.

Current Climate Policies

With the constant and seemingly insurmountable opposition to federal climate action in the US Congress, President Obama announced in his State of the Union speech in 2013 that he could wait no longer for gridlocked law-makers:

> I urge this Congress to get together, pursue a bipartisan, market-based solution to climate change, like the one John McCain and Joe Lieberman worked on together a few years ago. But if Congress won't act soon to protect future generations, I will. I will direct my Cabinet to come up with executive actions we can take, now and in the future, to reduce pollution, prepare our communities for the consequences of climate change, and speed the transition to more sustainable sources of energy (White House, 2013a).

A few months later, in June 2013, a Climate Action Plan was presented, outlining an approach to developing regulations for the largest GHG emitters in the United States (White House, 2013b). The plan has three main elements: to reduce domestic GHG emissions, to prepare the country for the impacts of climate change, and to be a leader in international climate change cooperation. In Obama's second term, therefore, new proposals on energy and climate policy increasingly focused on executive action that did not require congressional support.

A main element in the Climate Action Plan is to develop regulations to target carbon emissions from both new and existing power plants. Regulations are gradually being finalized with the objective of eventually regulating all major CO_2 emission sources in the United States. Most controversial is the Clean Power Plan (CPP) that was announced in June 2014, whereby the US Environmental Protection Agency (EPA) proposed a rule for implementing state-specific regulations to cut CO_2 emissions from existing power plants. The CPP assigns individual goals for emission cuts to each state, and the aggregate national target is to reduce emissions from power plants by 30 per cent by 2030 (EPA, 2014a). The regulations for new and existing power plants will disadvantage coal plants significantly more than they will natural gas plants because of the higher CO_2 content in coal.

The CPP has two main elements: state-specific goals for emission rate–based CO_2 reductions, and requirements for states to develop, submit and implement plans to reduce CO_2 emissions in their state. The EPA developed state-specific emission rate targets by, first, analysing the potential in each state for improvements in efficiency at carbon-intensive power plants; second, developing programmes that incentivize zero-emission or low-emitting energy sources; and third, developing energy efficiency programmes for end users such as homes and businesses (Federal Register, 2014). The current fuel mix, electricity market and other factors were considered when the EPA evaluated and decided upon a specific emission rate goal for each state that reflected each state's unique conditions. As a result, state emission rate goals vary from around 10 per cent to above 50 per cent (C2ES, 2014). State plans to meet the goals must be consistent with EPA guidelines. States can meet the targets alone, or can collaborate with other states on multi-state plans that could be more cost-effective and flexible, for instance cap-and-trade systems (Federal Register, 2014).

Several states have already enacted policy programmes aimed at reducing CO_2 emissions, often combined with phasing in more low-carbon energy sources in the electricity generation mix. In the Northeast, nine states have joined forces in the Regional Greenhouse Gas Initiative

(RGGI).[1] The aim is to cap and reduce CO_2 emissions from the power sector through an emissions trading system, which has operated since 2010. California implemented an emissions trading system in 2012 that covers a broader set of sectors in the state economy than the RGGI system. RGGI covers only the power sector. Moreover, as many as 27 states have enacted renewable portfolio standards of different strength and ambition levels, and states such as Texas and Colorado had significant growth in the share of renewable energy – in particular wind energy (C2ES, 2015).

Finally, EPA regulations developed since 2010 to reduce power plant emissions of toxic air pollutants such as mercury, ozone and fine particles would have an indirect effect as climate policy measures because they add to the pressure to retire old coal plants. For example, EPA's Cross-State Air Pollution Rule and the Mercury and Air Toxics Standards[2] require old coal-fired power plants to install new pollution control technology. These requirements give old coal-fired power plants an economic disadvantage, especially in comparison with natural gas-fired plants that are not affected by the regulations and also benefit from low natural gas prices. Consequently, these regulations can help accelerate the retirement of old coal-fired power plants and phase in more natural gas plants to replace them.

EXPLAINING THE US CLIMATE POLICY TRAJECTORY

President Obama's new approach to climate policy invites scrutiny concerning the political feasibility of building more acceptance for the CPP than was possible for climate legislation. This section gives an assessment of whether changes in fundamental parameters and driving forces are likely to increase the prospects for the CPP to be successfully implemented.

Fundamental Parameters

A first cut of the analysis is to look at how fundamental parameters affect policy-making processes. Both material and political parameters are crucial in this respect, and this section explores whether there have been significant changes in fundamental parameters that can help assess the likelihood of policy change. The analysis forms an important basis for understanding the dynamic between the contents of policy programmes proposed by the government and the shape and form of societal demand for policy change, which is explored in the next section.

Material parameters: from coal to shale gas

Domestic energy resources and their distribution across states have fundamentally affected federal climate policy processes because of the political preferences they create. Coal is an abundant energy resource in the United States, with mining activities generating jobs and income in more than 25 states (EIA, 2015a). Importantly, coal has a strong traditional role in producing low-priced electricity. For the United States in total, 39 per cent of electricity in 2014 was generated from coal, but states such as West Virginia, Missouri, Indiana, Kentucky, North Dakota, Ohio, Utah and Wyoming relied more than 80 per cent on coal for electricity generation (EIA, 2015b). In addition to abundant coal reserves, the United States has recently emerged as a major oil producer. The discovery and increasing production of large shale oil reserves changed the country's status from being a petroleum importer to becoming a petroleum exporter (EIA, 2014a). States such as Texas, Louisiana, Alaska, California and North Dakota have a large and growing petroleum industry that provides important resources, jobs and income. Moreover, new technologies for horizontal drilling and hydraulic fracturing have unlocked trillions of cubic feet of natural gas from shale formations in many states across the Midwest, South and Mountain West. The proven shale gas reserves are at least 348 trillion cubic feet according to EIA estimates (EIA, 2013a). Shale oil and shale gas resources have proven to be quickly producible in large volumes at relatively low cost, and have revolutionized US oil and natural gas production. Shale oil provided 29 per cent of total US crude oil production and shale gas provided 40 per cent of total US natural gas production in 2012 (EIA, 2014b). Hence, the shale oil and shale gas 'revolution' caused significant changes to material parameters in the United States. Such comprehensive changes happen rarely in the energy sector.

As a result, the US energy market is changing such that coal is losing its traditional place as the preferred energy source for power generation, and natural gas is increasingly taking coal's place. The vast increase in gas availability has lowered the price from a peak of \$13 per million British thermal units (MMBtu) in 2008 to approximately US\$3–4 per MMBtu in 2014 (EIA, 2014c). Because of the price drop and availability of gas, utility companies increasingly prefer investment in natural gas-fired power plants to coal-fired plants. Numerous retirements of old coal-fired power plants are now being planned (EIA, 2014d).

In the short term, a transition from coal to natural gas will reduce US GHG emissions because of natural gas's comparatively lower carbon density. Reports from the EIA show that CO_2 emissions dropped significantly since 2007, just before the shale gas transition started. Total US

CO_2 emissions in 2007 were 6023 million metric tons (MMT), and fell to 5290 MMT in 2012 bringing emissions down to 1994 levels (EIA, 2013b). A switch from coal to gas in the power sector was identified as one major cause of this reduction, along with other factors such as mild winters and reduced consumption during the economic crisis. Slight increases in energy-related CO_2 emissions were registered in 2013 because of greater coal use (EIA, 2014e).

Research shows that politicians representing states with large fossil fuel endowments will more likely vote against climate policies such as a cap-and-trade programme or low-carbon energy incentive structures that would negatively affect energy-intensive industries (Fisher, 2006; Skodvin, 2010). Fossil fuel resource endowments and the economic interests in states with such endowments create strong political preferences and advocacy coalitions in US climate and energy policy development. These preferences will likely remain stable and shape investment horizons for important state-level industries and infrastructure in ways that will affect federal policy debates. Hence, material parameters such as large fossil fuel endowments and related infrastructure investments have created status quo stability in US energy policy because these economic interests affect the configuration of political interests and the distribution of power and influence among US societal actors.

Alternative energy sources, such as wind and solar, have not yet been perceived by legislators as competitive with coal because of lower capacity, intermittence issues and price disadvantages (Victor, 2012). Although pressure groups such as environmental non-governmental organizations (ENGOs) and the American Wind Energy Association have formed advocacy coalitions and lobbied Congress for long-term subsidy structures to incentivize development of renewable energy technologies, their lobbying power has been ineffective (LaRussa, 2010). The renewable energy industry experienced unpredictable variations and cut-offs in the incentive structure over the past 20 years, which resulted in large variability in investment levels for renewable energy (Skodvin, 2010). Renewable energy use has grown substantially in some states – for example, Texas, Colorado and California – but no renewable energy source, technology or infrastructure can yet compete with natural gas or coal as an abundant, reliable, cheap energy source for electricity generation.

Continued reliance mainly on fossil fuels in electricity generation is likely despite mounting public opposition against 'fracking' in many states. For instance, in Colorado grassroots groups advocated to pose a 'ban on fracking' as a ballot question in the 2014 mid-term elections, but did not succeed (Energy Wire, 2014a). New York adopted a moratorium on fracking already in 2008, and Governor Cuomo made a decision in December 2014 to extend the moratorium indefinitely (Greenwire, 2014a). In some places where hydraulic fracturing has been going on for several years, local groups have formed to oppose further shale gas extraction because they fear harmful consequences. At least three concerns are linked to fracking. First, citizens living close by the fracking sites fear that chemicals used in the production process will leak into the groundwater. Fracking companies drill and inject fluid composed of water and a cocktail of chemicals into the ground at high pressure in order to fracture shale rocks and release natural gas. Another problematic side effect of fracking is that it might cause earthquakes. The most well-known case is that of a series of earthquakes near Youngstown, Ohio, in December 2011, where the largest quake had a magnitude of 4.0 on the Richter scale. The quakes were most likely caused by the underground injection of shale-drilling wastewater, as reported by Ohio government officials (Greenwire, 2012). A third consequence of fracking is increased methane emissions due to leakage during the production process. Calculating exactly how much methane is escaping during the fracking and processing of oil and gas is difficult. Estimates of methane emissions range from 1.5 per cent to 9 per cent (Brandt et al., 2014). Leakage happens in several parts of the production chain, and can be ameliorated somewhat by upgrading technology and introducing new regulations. According to a new study, leaks of methane from drilling sites in 2013 were 50 per cent higher than previously estimated by the EPA (ibid.).

In sum, despite some degree of public resistance to fracking, a high reliance on fossil fuels in electricity generation is likely to continue in the foreseeable future. The changes in material parameters in the US energy market have been significant and have led to emissions cuts because of a transition from coal to natural gas in the utility sector.

Institutional parameters

Two options for federal decision-making are common within the US political system. Decisions to change federal policy can be made either in a legislative process in the US Congress or by presidential decree where executive powers are invoked. Three governmental branches at the federal level share political power – the President (executive), Congress (legislative) and Supreme Court (judiciary). Federal climate and energy policy-making takes place in all of these political institutions. Legislation must have majority support in both chambers of Congress and be signed by the president to become law. A bill must pass many hurdles to be enacted: the House Subcommittee, the House Full Committee, the House

Rules Committee, the House, the Senate Subcommittee, the Senate Full Committee, the Senate, floor leaders in both chambers, the House–Senate Conference Committee, and the president. If the opposition wins at any stage, the bill fails (Hovi et al., 2012). In the implementation phase, the Supreme Court has the final power to interpret and decide in cases of dispute over federal law.

Because US federal political institutions work under rules for representation and decision-making that often result in high status quo stability, many barriers must be cleared before a bill can become domestic law. In other words, many veto players exist in the US system. Democrats and Republicans are often far apart ideologically, and the system's structure impedes rapid and significant changes to the status quo (Tsebelis, 2002). Indeed, few bills make it through the US legislative process.

During each stage in the congressional decision-making process, veto players and coalition builders have critical roles in blocking or passing a bill. Veto players in the energy and climate policy context are closely related to economically powerful fossil fuel energy interests. All politicians are concerned with their re-election prospects, and satisfying key constituent interests is their first priority (Arnold, 1990).

As a possible way of circumventing gridlock in Congress, US presidents have considerable freedom to pursue their own policy goals. Presidents have several unilateral tools at hand to pursue their policy agenda – including executive agreements, executive orders, presidential memoranda, proclamations and signing statements (Cooper, 2002). However, the president's authority also depends on political and institutional contexts. Executive action is less resistant to changes in political currents than legislative action is because a new president with a different political platform can reverse or remove it. Presidents must therefore consider how their opponents might respond to unilateral action and evaluate the likely degree of compliance, and the costs and benefits of acting unilaterally compared to instituting legislation in collaboration with Congress (Mayer, 2001; Shull, 2006).

Ever since Obama directed the EPA to design new regulations to cut GHG emissions from vehicles and power plants under the jurisdiction of the Clean Air Act, opponents in the US Congress have protested. One of the key criticisms is that Obama deprives law-makers of their right to be involved in major federal policy changes that have significant consequences for their constituents. Hence, Obama's executive approach to GHG regulations is controversial, and congressional opponents have introduced several bills that aim to remove the EPA's authority to regulate CO_2 emissions. These attempts have failed so far (Wallach, 2012). Under a new president with a different climate policy agenda, however, a presidential veto to protect EPA regulations is no longer guaranteed. Furthermore, a new president can decide to remove the EPA regulations altogether, even if it would require significant resources to stop such a regulatory process.

Driving Forces

The US policy-making process is also strongly influenced by driving forces that affect the prospects for implementation of the CPP. The following sections analyse ways in which governmental supply by the Obama administration and demand from various societal actors shaped policy-making.

Governmental supply

The role of the Obama administration as a driving force in climate policy has changed in Obama's second term. Although Obama often spoke in favour of climate legislation during his first term, he did little to push directly for action; for instance, he did not actively participate in coalition-building efforts in Congress. On the contrary, some assessments have shown that Obama failed to back Democratic legislators sufficiently at crucial junctures in the Senate law-making process in 2009–10 (Lizza, 2010). In 2013, at the beginning of his final term, Obama's focus shifted toward executive action, which gave the administration a much more active role as a pusher for climate policy change. Specifically, the CPP represents a much clearer ownership and commitment by the administration toward driving climate policy change. This regulation is aimed at limiting the use of coal, a goal made possible by the shale gas revolution that provided ample access to alternative, cheap and cleaner energy that can replace coal.

In addition to proposing stronger regulations that will cut CO_2 emissions, the Obama administration set up a tight timeline for public hearings on and finalization of CPP rules to improve the chances of implementation before Obama leaves office. The 2014 mid-term elections did, however, usher in the strongest Republican majority since 1928, and Republican leaders promised to make removal of Obama's CPP a priority for the 114th Congress (Greenwire, 2014b). Several tools are available for Congress to weaken the administration's regulatory capacity to finalize and implement the proposed regulations. For instance, Congress can make deep cuts in EPA's budget or insert riders in the national budget that prevent the administration from spending funds for specific climate policies.

The Obama administration also initiated new and stronger international climate policy to follow up promises in the Climate Action Plan to take international leadership to address climate change. In a joint statement in November 2014, President Obama and Chinese President Xi Jinping pledged new targets for emissions control in both countries. The United States pledged to cut emissions from 2005 levels by 26–28 per cent in 2025, while China pledged to cap its emissions by 2030 and phase in at least 20 per cent renewables in the energy mix within the same time frame. US pledges will be met through emissions cuts planned within the framework of the Climate Action Plan (White House, 2014). The credibility of US pledges to be a leader in international climate negotiations will therefore depend on the full implementation of the CPP and on additional emissions cuts.

The Obama administration's policy proposals were initiated with reference to scientific recommendations assembled by the US National Academy of Sciences and the IPCC, recommendations that resonate strongly among a large majority of Democrats. In 2014, as many as 88 per cent of Democrats believe that global warming is happening, and 82 per cent of Democrats support strict CO_2 emission limits on power plants (Leiserowitz et al., 2014). Generally, the majority of Democrats strongly believe in climate change science (McCright and Dunlap, 2011). Obama's push for climate policy change therefore had a clear mandate from the party. Nevertheless, Democrats representing states with a fossil fuel–intensive economy have formed a very vocal opposition to Obama's policy. In states such as West Virginia, a traditionally Democratic state with a large coal industry, Obama's climate policy initiatives have been very unpopular. Legislators from these states have tended to vote against the Democratic Party line in Congress, and oppose the president's executive action by supporting bills aimed at restraining EPA's authority to regulate CO_2 emissions. Clearly this split within the Democratic Party has weakened President Obama's ability to initiate changes in domestic climate policy through the legislative channel of policy-making, and has contributed to the administration's turn toward increased employment of executive action.

Societal demand

The general public's attitudes toward federal energy and climate policy are influenced by their levels of concern about energy security, energy prices and environmental protection. While about half of Americans (52 per cent) in 2011 believed that human actions are mostly to blame for climate change, their willingness to endure higher energy prices (for instance, as a result of carbon pricing) was strongly linked to the US economy's condition (Scruggs and Benegal, 2012). Polling trends spanning two decades show that when the US economy is suffering and unemployment is high, Americans tend to discourage policies that could increase energy costs. Conversely, in a booming economy, their willingness to pay for environmental protection is much higher (ibid.). These polling results suggest that public demand for low-carbon energy policy is partly linked to the economy's condition, and show that public support for federal climate policy action fell dramatically between 2008 and 2012 during the financial crisis.

In 2014, in step with economic recovery, public opinion had changed significantly. A report from Yale University indicated that 54 per cent of survey participants supported climate policy action (Leiserowitz et al., 2014). Specifically, support for EPA's CPP regulations was linked to the positive effect that power plant emissions reductions have for improving health conditions. Taking political party alignment into account, the survey showed that 72 per cent among Democrats and 46 per cent among Republicans supported the CPP (ibid.). Moreover, the survey found a clear difference between moderate Republicans and conservative Republicans. Moderate Republicans conveyed many of the same views as moderate and conservative Democrats did concerning the climate change issue, indicating that Republicans did not uniformly oppose climate policy. Liberal and moderate Republicans' views were close to the mainstream of American public opinion on climate change, while conservative Republicans' views stood out as distinctly different from those of the rest of the American public (ibid.). In general, public opinion also varies significantly between states with high fossil fuel energy dependence in the state economy, and states with weaker ties to fossil fuel energy (Bang and Skodvin, 2014).

Stakeholder and interest group attitudes to the CPP differed from sector to sector, often depending on whether their own economic activity would be negatively affected. Two main groups stood out: those that believed the economy would benefit from a transformation of the electricity generation system to eliminate carbon-intensive fuels, and those that believed that coal, oil and other fossil fuels should remain a part of the energy mix even in the future (Climate Wire, 2014a). The strongest negative reactions, as in previous rounds of climate policy debates, came from fossil fuel–dependent energy industries that risked higher costs. For instance, energy companies such as Southern Company and American Electric Power are among the biggest coal consumers in the United States, and together with other powerful energy firms they formed alliances with politicians opposed to Obama's policy proposals (Greenwire, 2014b).

Opposition to the CPP was organized in the legislative and judicial branches at both the federal and state levels of governance. In the US Congress, Republicans and law-makers representing fossil fuel–intensive states raised hard criticism of the CPP because it exposed the energy sector in their states. Many legislators were provoked by Obama's use of executive action because it represented a circumvention of Congress, and they expressed a clear request for involvement of the legislative branch in climate policy decisions since this policy field is of high national importance and new policy instruments would affect constituents directly (ibid.).

State-level politicians' reactions to the CPP varied from state to state, depending on the energy resource mix and which party had the majority. For example, Governor Rick Perry (Republican) of Texas released a statement just a few hours after the draft rule had become public, saying that the rule 'is the most direct assault yet on the energy providers that employ thousands of Americans' (Climate Wire, 2014b). The West Virginia legislature passed a law already in March 2014 – several months before the announcement of the CPP – that intended to make the state better positioned to sue the Obama administration over any regulations attempting to tighten pollution controls on power plants (*New York Times*, 2014). Another example of negative pressure is the Republican Attorneys General Association's very active engagement in organizing states' judicial efforts to stop Obama's regulatory agenda in the power sector (ibid.). In general, states with a clear fossil fuel energy profile and with a Republican governor or Republican majority in the state legislature protested sharply against the CPP.

Implementation of the CPP requires close interaction between the state and federal levels of government because states are given a key role in developing implementation plans, while the federal government will perform oversight. However, the individual interests of states as well as the multiple branches of government involved create ample opportunities for actors opposed to policy change to put up barriers. To be effective, governance at multiple levels should be of a reinforcing nature (Schreurs and Tiberghien, 2010). In a federal system, however, the strong role of the states can hinder policy implementation, especially since states are granted 'special solicitude' from the federal courts, a critical advantage that helps confer legal standing and means that a matter is less likely to be dismissed by the court. For the energy industry, therefore, states opposed to the CPP are extremely valued partners in trying to stop the Obama administration's regulatory agenda (*New York Times*, 2014).

In contrast, in states on the West Coast and in the Northeast, where energy resources are less fossil fuel intensive, the CPP was met with more enthusiasm. For instance, Governor Jerry Brown (Democrat) of California congratulated EPA for 'taking the next step on a path to not only reducing carbon pollution in the United States, but also in driving investment in new clean technology' (Climate Wire, 2014b). Others expressed approval of the Obama administration's leadership, and pointed out the need to act fast to address climate change. Environmentalists and ENGOs were among the strongest supporters of the CPP (ibid.).

Most state governors and regulators had more cautious reactions, reactions that to a certain extent reflect the complex energy landscape of the United States. Some states are already implementing various policies to upgrade their energy production systems, and many states have some form of incentive policy for renewable energy phase-in, for example, the 27 states that have Renewable Energy Portfolio standards. In presenting the rule, the EPA emphasized that state plans of compliance should build on existing policies and fortify existing programmes (EPA, 2014b). Some regulators decided to start working on plans and to gather information to develop well-informed recommendations for governors and legislatures (Energy Wire, 2014b).

In sum, societal demand and support for the CPP are mixed, even among state regulators who will be responsible for designing state-specific plans for meeting the reduction target. Strongly embedded differences of both an ideological and geographical nature play a role for how states' policy demands are shaped. The traditional cleavage between Republican and Democratic law-makers, especially at the federal level, seems to persist. The same goes for the traditional cleavage between states with fossil fuel–intensive economies and states that depend more on low emission or zero emission energy sources. Since the latter category of states is growing in step with the transition to natural gas in the electricity generation sector, the power relationship between these actors in the political debate may change in the future.

Supply–demand relationship

Many of the same issues that are central in US climate policy today have been debated for the past 20 years. Many of the political differences between domestic decision-makers persist too. By and large, it is Democrats and moderate Republicans that have proposed policy programmes aimed at changing domestic climate policy, while conservative Republicans have fought to sustain the status quo. Effective lobbying by powerful stakeholder groups in the energy sector has brought attention to the potential economic risks and costs related with strong climate policies, and especially the economic consequences of a carbon price for the coal industry and electricity prices. With the rise of the Tea Party, opposition to climate policy programmes increasingly turned into a Republican core issue, and caused more severe polarization between Republicans and Democrats on climate policy issues (Skocpol, 2013).

The institutional setting has constituted a complicated context because it stimulates the power of veto players and blocking coalitions. Deep reliance on fossil fuel sources for economic growth in many states constitutes the backbone in resistance to climate policy change. Political representatives from these states have been instrumental in blocking climate legislation in Congress, and Democrats from coal and fossil fuel–intensive states have contributed with key votes. In effect, the blocking votes by these veto-playing senators made it impossible for the Obama administration to pass climate legislation in Congress. Even when the Democrats enjoyed large majorities in the House and Senate in 2006–10, coal state Democrats aligned with Republicans to form a blocking coalition to stop climate policy change. This split within the Democratic Party contributed to bring about a change from legislative policy initiatives to executive action under the Obama administration.

In sum, the strongest societal demand in the United States is expressed as a robust and durable opposition to climate action, and a strong pressure to preserve the status quo. In comparison, the public pressure to act on the problem is much weaker. The Obama administration's CPP attempts to overcome distributional, geographical and ideological differences among policy-makers at the federal and state levels. The implementation process depends on the administration's ability to satisfy veto players sufficiently so they will contribute in developing and implementing CPP state plans.

LIKELY POLICY TRAJECTORY

President Obama's new climate policy initiatives – the Climate Action Plan and the CPP – were explicitly intended to be a way to circumvent veto players in Congress, and to implement climate policy regulations despite deep cleavages among federal legislators. The strategy was to turn to the state level and engage state legislators and regulators to develop regulations specifically designed to handle local circumstances.

The CPP is set to be implemented before Obama leaves office in 2017. For the implementation phase, two scenarios are likely. In the first scenario, we will see a development where state-level action spurs changes in the dynamics of societal demand for climate policy. The process of developing state plans may become a knowledge-building process for state regulators, making them aware of various possibilities to reduce emissions without intolerable costs. Expanding on previously introduced state programmes – such as a Renewable Portfolio Standard – is one realistic route, for instance. Obama's federal initiative can then provoke state-level engagement and policy development, resulting in state-level experience eventually spreading to the federal level where it will change the preference structure of federal-level law-makers. Enactment of federal climate legislation would then be possible. Scholars analysing climate policy development in the United States and in the European Union have previously described such processes as multilevel reinforcement, boomerang federalism, or the role of states as laboratories of federal policy (Rabe, 2008; Schreurs and Tiberghien, 2010; Fisher, 2013).

Alternatively, in the second scenario, we will see a development whereby state legislators and regulators fight the CPP every step of the way through the state and federal legislatures, and in the courts. Implementation can be delayed in several ways. First, states have the right to request extra time (up to three years under special circumstances) to prepare plans for meeting the reduction target (Federal Register, 2014). Furthermore, states can initiate court cases against the EPA if they can prove that regulations will unfairly harm the state and/or its economy. Several coal states are already planning such a lawsuit against the EPA (Energy Wire, 2014b). The Constitution allows the judicial branch final decision-making power in disputes about the level and force with which laws are to be implemented. Third, a state's delegation to the US Congress can attempt to build coalitions to pass legislation that could remove the EPA's authority to regulate CO_2 emissions. Many such attempts have been made against the EPA under Obama's presidency, and more are likely. However, as long as the presidency is in the Democrats' hands, it is unlikely that such legislation will pass *and* avoid a presidential veto.

The next US president can in theory remove EPA regulations with a stroke of the pen once in office in 2017, although in practice it would be resource demanding to stop a regulatory process in motion. The Obama administration is therefore in a hurry to implement as much as possible of the CPP before it leaves office, and in as many states as possible. If state plans to fulfil the CPP have been implemented in most states by 2017, and if significant resources have been invested in this work, the next president will find it harder to remove the plan altogether.

Material parameters such as distribution of energy resource endowments across states and this distribution's consequences for state economies will likely continue to play a major role in climate and energy policy developments in the United States. Similarly, how states' interests are manifested in federal political institutions will remain crucial in determining the political feasibility of policy alternatives. However, the political consequences of the shift from coal to natural gas in the utility sector could be significant. US energy-related CO_2 emissions were reduced drastically in recent years as a result of the shift, reaching 1994 levels in 2012 (EIA, 2013b). Reductions can be expected to continue if the new regulations are implemented. If not, a rise in natural gas prices could result in a transition back to coal, as indeed happened to a certain extent in 2013 when energy-related CO_2 emissions increased slightly compared to 2012 because of more coal consumption in the energy sector (EIA, 2014e).

The combined effect of low natural gas prices and regulations to reduce the use of coal in power generation can have a big impact on power relationships in the US energy market and in energy politics. A major and sustained shift to natural gas in the utility sector could change the power relationship by splitting the interests of the coal industry and the natural gas industry, which previously have acted together in questions related to climate and energy policy. Changed power structures along these lines could have a major impact in future policy processes.

As the US economy recovers from the economic crisis, stronger public demand for a more comprehensive response to the climate change problem is likely, especially if negative consequences of global warming become more evident. Experience from previous environmental policy debates suggests, however, that such demand is feeble and cyclical (Downs, 1972). Positive experience with state regulations could change this dynamic, as mentioned. But only if public concern coalesces with the interests of vital industries and strong interest groups to demand new policy solutions that are attractive both economically and environmentally can we expect stable changes to federal US energy and climate policies.

The implementation of the CPP is also crucial for international climate cooperation efforts, because the United States must be able to deliver on its bilateral deal with China. Without full implementation of the CPP, the United States has no credible domestic policy plan for realizing its 2025 emissions reduction target.

NOTES

* The author would like to thank Tora Skodvin, Arild Underdal and Steinar Andresen for valuable comments to previous versions of this chapter.

1. Member states in RGGI are Rhode Island, New York, Vermont, Connecticut, Maine, New Hampshire, Delaware, Maryland and Massachusetts.
2. The Cross-State Air Pollution Rule requires states to significantly improve air quality by reducing power plant emissions that contribute to ozone and/or fine particle pollution in other states. The Mercury and Air Toxics Standards require power plants to limit their emissions of toxic air pollutants like mercury, arsenic and metals, and set technology-based emissions limitation standards for mercury and other toxic air pollutants, reflecting levels achieved by the best-performing sources currently in operation (EPA, 2014b).

REFERENCES

Arnold, R.D. (1990), *The Logic of Congressional Action*, New Haven, CT: Yale University Press.

Bang, G. (2010), 'Energy security and climate change concerns: triggers for energy policy change in the United States?', *Energy Policy*, **38**(4), 1645–53.

Bang, G. and T. Skodvin (2014), 'US climate policy and the shale gas revolution', in T. Cherry, J. Hovi and D. McEvoy (eds), *Toward a New Climate Agreement: Conflict, Resolution and Governance*, London: Routledge.

Bang, G. and M.A. Schreurs (2010), 'A green new deal: framing US climate leadership', in R. Würzel and J.M. Connelly (eds), *The European Union as a Leader in International Climate Change Politics*, London: Routledge, pp. 235–51.

Brandt, A.R., G.A. Heath, E.A. Kort, F. O'Sullivan, G. Petron and S.M. Jordaan et al. (2014), 'Methane leaks from North American natural gas systems', *Science*, **343**(6172), 733–5.

Climate Wire (2014a), 'A full spectrum of reactions to EPA power plant proposal from corporate groups', 3 June, accessed 26 January 2015 at http://www.eenews.net/climatewire/stories/1060000578/search?keyword=EPA+rule (subscription required).

Climate Wire (2014b), 'Proposed EPA power plant rule launches a new experiment in federalism', 3 June, accessed 26 January 2015 at http://www.eenews.net/climatewire/stories/1060000595/search?keyword=EPA+rule.

Cooper, P.J. (2002), *By Order of the President: The Use and Abuse of Executive Direct Action*, Lawrence, KS: University of Kansas Press.

C2ES (2014), 'Carbon pollution standards map, proposed state emissions rate targets', accessed 12 December 2014 at http://www.c2es.org/federal/executive/epa/carbon-pollution-standards-map.

C2ES (2015), 'US states and regions – climate action', accessed 25 January 2015 at http://www.c2es.org/us-states-regions.

Downs, A. (1972), 'Up and down with ecology – the issue-attention cycle', *Public Interest*, **28**(Summer), 38–50.

Energy Wire (2014a), 'Industry commits nearly $2M to fight state anti-fracking ballots', 27 March, accessed 31 March 2014 at http://www.eenews.net/energywire/stories/1059996803/search (subscription required).

Energy Wire (2014b), 'Behind the noise, central states study EPA rule cooperation', accessed 26 January 2015 at http://www.eenews.net/energywire/2014/12/03/stories/1060009833 (subscription required).

Federal Register (2014), *Carbon Pollution Emission Guidelines for Existing Stationary Sources: Electric Utility Generating Units*, 18 June, accessed 12 December 2014 at https://www.federalregister.gov/articles/2014/06/18/2014-13726/carbon-pollution-emission-guidelines-for-existing-stationary-sources-electric-utility-generating#h-9.

Fisher, D.R. (2004), *National Governance and the Global Climate Change Regime*, Lanham, MD: Rowman & Littlefield Publishers.

Fisher, D.R. (2006), 'Bringing the material back in: understanding the US position on climate change', *Sociological Forum*, **21**(3), 467–94.

Fisher, D.R. (2013), 'Understanding the relationship between sub-national and national climate change politics in the United States: toward a theory of boomerang federalism', *Environment & Planning C: Government & Policy*, **31**(5), 769–84.

Greenwire (2012), 'Wastewater injection well sparked earthquake – Ohio officials', 9 March, accessed 31 March 2014 at http://www.eenews.net/greenwire/stories/1059961197.

Greenwire (2014a), 'Cuomo admin bans fracking in N.Y.'s Marcellus Shale', 17 December, accessed 27 May 2015 at http://www.eenews.net/greenwire/stories/1060010734/search?keyword=New+York+moratorium.

Greenwire (2014b), 'McConnell, Inhofe vow to strike EPA power proposal', 3 June, accessed 26 January 2015 at http://www.eenews.net/greenwire/stories/1060000633/search?keyword=epa+power+plant+rule.

Hovi, J., D. Sprinz and G. Bang (2012), 'Why the United States did not become a party to the Kyoto Protocol: German, Norwegian and US perspectives', *European Journal of International Relations*, **18**(1), 129–50.

LaRussa, C. (2010), 'Solar, wind power groups becoming prominent Washington lobbying forces after years of relative obscurity', *Open Secrets Blog*, 30 March, accessed 26 January 2015 at http://www.opensecrets.org/news/2010/03/solar-wind-power-becoming-prominent/.

Leiserowitz, A., E. Maibach, C. Roser-Renouf, G. Feinberg and S. Rosenthal (2014), *Climate Change in the American Mind*, Yale University Project on Climate Change Communication, October, accessed 26 January 2015 at http://environment.yale.edu/climate-communication/files/Climate-change-American-Mind-October-2014.pdf.

Lizza, R. (2010), 'As the world burns', *The New Yorker*, 11 October.

Mayer, K. (2001), *With the Stroke of a Pen: Executive Orders and Presidential Power*, Princeton, NJ: Princeton University Press.

McCright, A.M. and R.E. Dunlap (2011), 'The politicization of climate change and polarization in the American public's views of global warming, 2001–2010', *The Sociological Quarterly*, **52**(2), 155–94.

New York Times (2014), 'Energy firms in secretive alliance with Attorneys General', 6 December, accessed 7 December 2014 at http://www.nytimes.com/2014/12/07/us/politics/energy-firms-in-secretive-alliance-with-attorneys-general.html?hp&action=click&pgtype=Homepage&module=first-column-region&region=top-news&WT.nav=top-news&_r=1.

Rabe, B. (2008), 'States on steroids: the intergovernmental odyssey of American climate policy', *Review of Policy Research*, **25**(2), 105–28.

Schreurs, M. and Y. Tiberghien (2010), 'European Union leadership in climate change: mitigation through multilevel reinforcement', in K. Harrison and L.M. Sundstrom (eds), *Global Commons, Domestic Decisions: The Comparative Politics of Climate Change*, Cambridge, MA: MIT Press, pp. 23–66.

Scruggs, L. and S. Benegal (2012), 'Declining public concern about climate change: can we blame the great recession?', *Global Environmental Change*, **22**(2), 505–15.

Shull, S.A. (2006), *Policy by Other Means: Alternative Adoption by Presidents*, College Station, TX: Texas A&M University Press.

Skocpol, T. (2013), 'Naming the problem: what it will take to counter extremism and engage Americans in the fight against global warming', paper presented at the symposium on 'The Politics of America's Fight Against Global Warming', Harvard University, 14 February.

Skodvin, T. (2010), 'Pivotal politics in US energy and climate legislation', *Energy Policy*, **38**(8), 4214–23.

Tsebelis, G. (2002), *Veto Players. How Political Institutions Work*, Princeton, NJ: Russell Sage Foundation and Princeton University Press.

US Energy Information Administration (EIA) (2013a), 'Proved reserves of natural gas (trillion cubic feet)', accessed 1 May 2015 at http://www.eia.gov/cfapps/ipdbproject/IEDIndex3.cfm?tid=3&pid=3&aid=6.

US Energy Information Administration (EIA) (2013b), 'US energy-related carbon dioxide emissions, 2012', accessed 25 January 2015 at http://www.eia.gov/environment/emissions/carbon/archive/2012/index.cfm.

US Energy Information Administration (EIA) (2014a), 'Outlook for US shale oil and gas', presentation given at Argus Americas Crude Summit, 22 January by A. Sieminski, EIA Administrator, accessed 25 January 2015 at http://www.eia.gov/pressroom/presentations/sieminski_01222014.pdf.

US Energy Information Administration (EIA) (2014b), 'Shale oil and shale gas resources are globally abundant', *Today in Energy*, 2 January, accessed 25 January 2015 at http://www.eia.gov/todayinenergy/detail.cfm?id=14431.

US Energy Information Administration (EIA) (2014c), 'Natural gas weekly update', accessed 25 January 2015 at http://www.eia.gov/naturalgas/weekly/archive/2014/09_11/index.cfm.

US Energy Information Administration (EIA) (2014d), 'Planned coal-fired power plant retirements continue to increase', *Today in Energy*, 20 March, accessed 31 March 2014 at http://www.eia.gov/todayinenergy/detail.cfm?id=15491.

US Energy Information Administration (EIA) (2014e), 'US energy-related CO_2 emissions in 2013 expected to be 2% higher than in 2012', *Today in Energy*, 13 January, accessed 31 March 2014 at http://www.eia.gov/todayinenergy/detail.cfm?id=14571.

US Energy Information Administration (EIA) (2015a), 'Coal data browser: aggregate coal mine production 2012', accessed 25 January 2015 at http://www.eia.gov/beta/coal/data/browser/.

US Energy Information Administration (EIA) (2015b), 'State profiles and energy estimates', accessed 25 January 2015 at http://www.eia.gov/state/.

US Environmental Protection Agency (EPA) (2014a), 'Clean Power Plan proposed rule', accessed 25 January 2015 at http://www2.epa.gov/carbon-pollution-standards/clean-power-plan-proposed-rule.

US Environmental Protection Agency (EPA) (2014b), 'Air and radiation – regulations and statutes', accessed 25 February 2015 at http://www.epa.gov/air/oarregul.html.

Victor, D. (2012), *White Paper on Energy Security and Global Warming from the World Economic Forum's Global Agenda on Energy Security*, accessed 24 February 2014 at http://www3.weforum.org/docs/WEF_GAC_WhitePaper EnereySecurityGlobalWarming_2012.pdf.

Wallach, P.A. (2012), *US Regulation of Greenhouse Gas Emissions*, Brookings Institution Report, October, accessed 1 May 2015 at http://www.brookings.edu/~/media/research/files/papers/2012/10/26%20climate%20change%20wallach/26%20climate%20change%20wallach.pdf.

White House (2013a), 'Remarks by the President in the State of the Union Address', US Capitol, Washington, DC, 12 February, accessed 14 December 2014 at http://www.whitehouse.gov/the-press-office/2013/02/12/remarks-president-state-union-address.

White House (2013b), 'Climate change and President Obama's Action Plan', June, accessed 25 January 2015 at http://www.whitehouse.gov/climate change#section-carbon-pollution.

White House (2014), 'US–China joint announcement on climate change and clean energy cooperation', Fact sheet, 11 November, accessed 26 January 2015 at http://www.whitehouse.gov/the-press-office/2014/11/11/fact-sheet-us-china-joint-announcement-climate change-and-clean-energy-c.

World Bank (2015a), 'United States', accessed 6 May 2015 at http://data.worldbank.org/country/united-states.

World Bank (2015b), 'Access to electricity (% of population)', accessed 6 May 2015 at http://data.worldbank.org/indicator/EG.ELC.ACCS.ZS.

9. Comparative analysis and conclusions

Guri Bang, Arild Underdal and Steinar Andresen*

INTRODUCTION

The impact of the new climate agreement negotiated in Paris will largely depend on the domestic policies of the key actors studied in this volume. All of them are attuned to addressing the climate change problem but the levels of commitment and effort vary, as the case studies have revealed. In our analysis we have aimed at improving the understanding of what drives and what blocks climate change mitigation policies. First, we have regarded material and political parameters as crucial in terms of defining the scope of policy change. Energy resource endowments, infrastructure investments and other material factors severely limit a country's policy options. Institutional political structures restrain fundamental and rapid policy changes because they regulate who has voice in the political process and how interaction between political adversaries takes place.

Second, within the political feasibility scope provided by material and political parameters, the interplay between these parameters and principal driving forces, notably governmental climate policy supply and societal demand for such policy, is decisive for the direction and force of policy change. Our analysis shows that when we focus on governmental supply factors, we add layers of understanding and knowledge about the direction and scope for policy change within each of the actors. For instance, the analysis of governmental supply of new climate policy in the United States under the Obama administration shows that material and institutional parameters put severe restraints on policy change options, but also that these parameters provide the US president with executive powers to introduce new policy proposals – powers that President Obama has increasingly exploited. Adding a focus on societal demand provides additional insight into the strength of advocacy coalitions that press for policy change and of veto players preferring the status quo.

When we analyse the interplay between both of these factors – fundamental parameters and driving forces – we get a comprehensive picture of the drivers and barriers in each actor's climate policy and the prospects for change within the short to medium term. Hence, by first defining basic parameters that set the scope for energy use and policy interaction, and then adding layers of more complex information about the mechanisms and conditions for governmental supply and societal demand, we are able to assess policy trajectories and the prospects for policy change.

An important underlying assumption is that, although in part reflecting fundamental material parameters such as energy resource endowments, *government policies* play significant roles in shaping countries' GHG emission practices and trajectories. Before assessing and comparing the main findings from our case studies, a critical look at that assumption seems appropriate.

TESTING THE 'NULL HYPOTHESIS'

In the introductory chapter we defined the I = PAT formula as our null hypothesis. The formula conceives of the aggregate human imprint on the environment as a multiplicative function of population, affluence and technology. We now proceed to assess the explanatory power of this formula by exploring how well it can account for *changes* in the GHG emissions of our seven key actors between 1990 and 2011. In line with the theoretical reasoning behind the formula, we treat aggregate impact (I) – the product of the three components – as the principal dependent variable. Impact is measured in terms of emissions of all GHGs as well as in terms of emissions of CO_2 only. While population and affluence (GDP per capita) are supposed to increase impact, technology can work both ways, rendering the choice of operational indicator(s) important. Here we run the analysis with technology measured in terms of energy efficiency, more precisely CO_2 emissions per unit of GDP. This choice is premised in part on our interest in aligning with previous studies, and in part on the fact that for all the alternatives we considered comparable time series data for our sample of actors were hard to find. We do recognize that this indicator is 'biased' in the sense that it represents a positive abatement effect. Technological development will usually also enhance productivity and incomes and thus indirectly contribute to higher emissions, but that effect is already partly represented by the affluence factor.

Overall, the predictions derived from the I = PAT formula correlate positively with our observations. For the world at large, bivariate correlations are 0.980*** when all GHGs are included and 0.981*** when we focus on CO_2 emissions only.[1] For our sample of key actors, the corresponding coefficients are much lower, 0.256** and 0.185* respectively. Zooming in on each individual actor, nine out of 14 correlations are positive and significant at the 0.001 level. Moreover, at least for our sample, each of the three driving forces seems to work in the direction expected even when we 'disconnect' an individual driving force from the formula and control for the other two (Table 9.1). For the 1990–2011 period, increased energy efficiency (here representing the technology factor) offsets the impact of population growth for all actors except Brazil and India. However, the *combined* effect of population growth and increased affluence overwhelms improvements in energy efficiency for all seven actors and for the world at large. Given projections of further population growth and the political imperative of eradicating poverty and improving living conditions more generally, this is bad news for mitigation. *Decoupling* GHG emissions from the combined PAT effect is a necessary condition for progress in mitigating the impact of human activities on the global climate system.[2]

Table 9.1 The relative impacts of population, affluence and technology

	Emissions, all GHGs		Emissions, CO_2 Only	
	Beta (standardized)	*t*-test	Beta (standardized)	*t*-test
Population	0.376***	4.256	0.448***	5.811
Affluence	0.493***	6.317	0.670***	9.817
Technology	–0.372***	–4.933	–0.320***	–4.850
Constant	(4892.161)***	4.636	(2859.200)**	3.400
Adjusted R^2	0.488		0.609	

Note: Linear regression, our sample [data are available at the authors' project website upon request: http://www.cicep.uio.no/english/], $N = 151$. Partial correlation analysis yields similar results.

On closer inspection, however, we find some striking differences in the formula's performance across countries. In two cases – China and India –

we observe almost perfect (linear) fit: correlations = 0.996*** and 0.992*** respectively, all GHGs included. For Brazil, Russia and the European Union the corresponding coefficients are –0.058, –0.140 and –0.567*** respectively.[3] Moreover, for Brazil, the European Union and Japan the performance of the I = PAT formula improves when we focus on CO_2 emissions only. The contrast is particularly striking in the case of Brazil, where correspondence between formula predictions and observed emissions rises to 0.953*** when only CO_2 emissions are included. A plausible interpretation of such 'deviant' cases is that emission trajectories are heavily influenced by factors not included in the formula. For example, in the case of Brazil, aggregate GHG emissions increased largely as predicted by the formula from 1990 to 2005 (0.720***) but then declined by nearly 40 per cent over the next six years. This reversal seems largely policy driven, because the Brazilian government, with parliamentary and international support, began implementing effective policies to reduce deforestation.[4] In the case of Russia, the collapse of the Soviet Union, followed by a difficult and erratic transition towards a more democratic and market-based system, led its GDP per capita to fall by about 30 per cent between 1990 and 1998, followed by an increase of more than 300 per cent between 1998 and 2011. Its population declined slowly but steadily from 1993 to 2009. The carbon intensity of its economy at first increased and then declined as old factories and other infrastructural components were upgraded or replaced. Turbulence of such magnitude and complexity seems to overwhelm the I = PAT formula and to call for an approach that includes political and economic institutions and processes in the equation.

Also worth noting in the cases of Brazil, the European Union, Japan and the United States is the recent 'decoupling' of GHG emissions from the I = PAT formula. Throughout the 2005–11 period, a 1 per cent (average) increase in I = PAT scores is associated with GHG emission *reductions* on the order of 0.8, 0.8, 0.5 and 0.3 per cent, respectively. Over the same period, a 1 per cent increase in I = PAT scores for China and India is associated with a near-equal *increase* in GHG emissions (0.9 per cent). Except for Brazil, the difference between these two groups is broadly consistent with the environmental Kuznets curve hypothesis. A good explanation will likely have to include factors other than high income levels, but given aggregate world trends of increasing GHG emissions, any instance of decoupling from the I = PAT formula merits attention.

For our study, the take-home message may be summarized as follows. The I = PAT formula identifies and combines important anthropogenic sources of environmental change and can help us predict and understand main patterns, at least at the macro level. Over the 21-year period included in this analysis, the formula's predictions of aggregate world GHG emissions show almost perfect fit with observed trajectories. At the level of individual countries, however, results are far more diverse. Most importantly, we observe some recent instances of decoupling of emission trajectories from the I = PAT formula. Such 'deviance' cannot be explained by that formula itself. The conceptual framework applied in this study offers one plausible alternative – highlighting domestic politics and government policies. Brazil's cut of GHG emissions by nearly 40 per cent during no more than six years shows that in certain settings policy change can achieve remarkable results, at least temporarily. To understand how significant change can be initiated and governed, we need a model that explicitly includes policies and politics in the equation. The I = PAT formula fails to meet that requirement. For our purpose, in-depth comparative analysis of actor- and context-specific factors is needed – particularly if the ambition is to understand developments sufficiently well to offer informed advice to policy-makers and stakeholders. This is exactly the kind of analysis we engage in in this volume.

EXPLANATORY PATTERNS

Changes in policy trajectories occurred in all of the seven key actors over the past decade, and our case studies show that in explaining these changes we find distinctive policy-related variation across the cases. Clearly, policy-making took place in distinct institutional and structural contexts that led into different pathways, and the case studies describe these differences. Still, some noteworthy *shared patterns* emerge. At a very general level, we find that policy change largely depends on economic and technological conditions in each case as well as on structural parameters like domestic energy resource endowments. More specifically, we found that alignment of economic opportunity, technological breakthrough and energy security concerns created an important cluster of factors for triggering policy changes. However, we also found that major policy change requires strong political push from the government in combination with acceptance by and support from key societal actors to be politically feasible. This chapter will go into more detail on some common explanatory features and some key *differences* in domestic climate policy decision-making processes across cases. We review case study findings in light of the assumptions formulated in the introductory chapter.

The first assumption we made was that material and political parameters normally have a stabilizing effect on policies, and will account for a significant proportion of the variance found in our sample. In other words, we expected material as well as institutional parameters to be profoundly important explanatory variables. The second assumption we made was that policy development must be understood as driven by two generic forces: demand and supply. These forces interact and co-produce policies, but the ways they do so vary significantly, depending on the characteristics of political institutions, social cultures and other country-specific features.

We specified these general assumptions to identify the expected constraining and facilitating effects of *material* and *political* parameters, and to identify the strength and form of *governmental supply* and *societal demand* as driving forces for change. Our case study evidence confirms that material and political parameters fundamentally define the scope and basis for policy choices, while the strength and form of supply and demand help explain the ambitiousness and feasibility of policy change within those limits.

Material Parameters

In Chapter 1, we assumed that domestic energy resource endowments and related concerns about domestic energy production, distribution and consumption would form an important cluster of explanatory factors for climate policy trajectories. We expected that energy needs in the electricity sector would play a particularly important role for policy-makers, and that the level of mitigation costs would affect decisions, especially in the energy sector. The more expensive policy measures are, the more they negatively affect energy prices, and the more they hinder the use of domestic energy resources, the less likely they are to be enacted. We assumed that this effect would also be linked to the degree of fossil fuel dependency each actor experiences, because fossil fuels traditionally have been cheaper than low-carbon or carbon-free energy sources, and because reliance on established infrastructure will often be cheaper in the short run.

In the case studies we find strong support for this assumption. We find that domestic energy resources, and especially the degree of fossil fuel dependency, are important explanatory factors for policy outcome in all seven actors (Table 9.2). In our sample, energy efficiency is negatively although weakly correlated with domestic energy resource endowments (Spearman = –0.26). China, India and the United States are the most deeply coal-dependent actors, with China producing 70 per cent of its electricity in coal plants, India 59 per cent and the United States 39 per cent. While the European Union and Japan are less dependent on coal, their energy security is tightly linked to their high levels of fossil fuel imports – especially natural gas. Natural gas also forms a large part of Russia's energy consumption, and Russia has abundant domestic reserves. The economy is very fossil fuel dependent, with 25 per cent of GDP linked to the petroleum sector. Brazil is the country in our sample least dependent on fossil fuel, although this may change because of a quickly developing oil industry.

Table 9.2 Relative energy resource endowments per capita, with the United States as baseline (= 1.00)

Actor	Coal	Natural Gas	Renewables
USA	1.00	1.00	1.00
EU	0.15	0.14	0.27
Japan	0.00	0.01	0.08
Russia	1.47	11.08	1.82*
Brazil	0.04	0.07	0.57*
China	0.11	0.10	0.21
India	0.07	0.03	0.08

Notes:

For fossil fuels, figures refer to 'proved reserves' as reported by the US Energy Information Administration. International Energy Statistics for 2011 (coal) and 2012 (natural gas). For renewables we have used estimates of 'minimum technical potentials' and included four major sources: solar (weight 0.4), wind (0.2), bio (0.2) and hydro (0.2).

All coefficients refer to *per capita* endowments, based on population statistics for 2012.

* Indicates higher-than-normal uncertainty pertaining to one or more of the estimates for renewables.

Sources: Solar, Timilsina et al. (2012); wind, Lu et al. (2009); bio, Smeets et al. (2007) and Batidzirai et al. (2012); hydro, World Energy Council (2010).

A dominant finding across all case studies is that concerns about energy security, demand for low energy prices and continued economic growth are closely connected and very strong explanatory factors. Policy-makers are extremely attentive to these issues, and factor in such concerns when making climate and low-carbon energy policy decisions. Moreover, political influence seems to reflect the current centrality of different types of energy sources. For instance, coal has for decades been counted as a reliable and cheap energy resource that is crucial for continued economic growth, and coal has a central place in energy policy planning by most of our actors. Thus, energy sources that already have a central place in the energy mix are difficult to replace because they are deeply embedded in the energy system. This lock-in effect applies to jobs and infrastructure investments related to the extraction, transport, combustion and use of traditional fossil fuel energy sources.

There are, however, some crucial differences across cases in terms of how the energy resource mix influences policy trajectories. As shown in Table 9.2, actors' domestic energy resource endowments vary hugely. One way that this variation manifests itself is that energy security concerns are closely related to the degree of fossil fuel import dependency. Our case studies show that actors with high import dependency are open to energy sector reform that includes low-carbon energy growth. Both the European Union and Japan have invested in fossil-free energy sources to compensate for lack of domestic fossil fuel reserves. While the European Union has expanded its share of renewables substantially, Japan for long relied heavily on nuclear energy. Before the Fukushima accident in 2011, Japan's emissions were on a downward trend. But when nuclear power plants were shut down as a reaction to the catastrophe, they were replaced by imports of fossil energy. Japan's emissions are now increasing. As a result, energy security has become a more prominent issue for Japan after Fukushima, and will likely lead to both re-opening of (some of) the nuclear plants and more investment in renewable energy. The European Union also experienced energy security as a problem because most EU member states are energy importers. Energy security concerns amplified when Russia threatened to use energy supply as a political weapon in the Ukraine crisis. In EU countries with high import dependency, like Germany, energy security concerns helped open a window for a transition to a green energy system and thus helped push climate policy processes forward. In contrast, other EU member states, such as Poland and the Czech Republic, have significant domestic coal reserves and are less willing to take part in the green energy transition. Brazil is another example where energy security concerns caused investment in low-carbon energy alternatives, namely a massive development of home-grown ethanol fuel for the transport sector to avoid oil import dependency.

Going further into the effect of energy resource mix on policy trajectories, we see that technological development can have important effects on GHG emissions as well as on climate policy trajectories. For instance, technology changes in the US energy sector made available hydraulic fracturing technology that gave access to huge natural gas reserves. Because of this energy market change, cheap natural gas is increasingly replacing coal in US power plants, which was unthinkable as recently as 2009. The shale gas boom has changed the picture for the United States, since natural gas can now compete with coal in terms of market price and available supplies. The United States is also among the world's top three investors in renewable energy, but compared to investments in domestic fossil fuels, these investments are small, and so far renewable energy sources amount to only about 10 per cent of the domestic energy mix. In China, development of solar panel technology at low cost facilitates the phase-in of more renewable energy in China's energy mix. China's massive investments in renewables will to some extent reduce its heavy dependence on coal, although renewables account for only about 10 per cent of the energy mix in China so far. In addition, natural gas is expected to have a crucial role in the future Chinese energy mix. According to the new energy plan, China will have 15 per cent non-fossil energy by 2020, but also at least 10 per cent natural gas (Bo, 2014). China is importing most of the natural gas, but is also planning to extract what natural gas reserves it has, conventional and unconventional. The emphasis on natural gas is fairly recent (from 2014), and linked to mitigation of air pollution. In cities, natural gas will continue to replace coal as the main household fuel, already an ongoing trend.

A third important finding related to material parameters is that all of the seven actors consider a stable supply of low-priced energy to be important for economic development. This attitude works as a barrier to ambitious climate policy in several of the cases we studied. As mentioned, the interplay between energy resource endowments and appropriate energy technology can significantly constrain or expand the scope of politically feasible climate policy options. Moreover, the more reliant a country is on domestic fossil fuels, the more influential energy sector concerns are for policy-making. In several of the cases that we studied, we found a relationship between the economic role of fossil energy and the power of energy sector actors in setting priorities for climate and energy policy. In Russia and the United States, for example, we found that oil and coal sector interest groups, respectively, were powerful and dominant because of the crucial economic interests that they represent. In effect, preferences shaped by energy resource endowments establish important premises limiting the scope for policy change. In the United States and China domestic access to cheap coal has been a very important parameter constraining the political feasibility of ambitious climate policies in both countries. The same story is dominant in India, where coal, domestic and imported, is a pillar of economic growth, and the outlook for replacement by renewables is bleak. Rather, expansion of renewables happens in parallel with increased fossil energy use in India, China and the United States.

In conclusion, our assumptions regarding the effect of material parameters were supported by the case study findings in three ways. First, as expected, an actor's endowment of domestic energy resources and (especially) the degree of fossil fuel dependency were important explanatory factors for climate policy trajectories. Second, energy security concerns significantly affected whether politicians prioritized renewable energy investments. Third, the more reliant a country is on domestic fossil fuels, the more influential energy sector concerns are for policy-making and the more powerful energy sector actors are in setting priorities for climate and energy policy.

Political Parameters

We expected political systems to differ with respect to the degree of decentralization along the vertical axis and the degree of fragmentation along the horizontal axis, and we assumed that both of these variables would affect the prospects for climate policy change. Specifically, we expected that more centralization of power would facilitate policy change, whereas political systems with a high degree of multilevel governance would have more veto points preserving the status quo. In Chapter 1, we also assumed that because democracies have a much wider range of channels and opportunities for societal actors to express demand and support than autocracies have, more societal actors would be involved in political processes in democracies and that policy-makers would have stronger incentives to pay attention to their voters' concerns.

The case studies confirm that political systems differ largely as expected (Table 9.3). In a centralized political structure like China's, important decisions are made at the top level and transferred down through the decision-making hierarchy to be implemented at the regional and local levels of government. There is more room in the implementation phase for regional- and local-level delay or adjustment of policies. In policy-making processes, a few key institutions in China play crucial roles in determining the direction and ambition level of climate policy proposals. Russia also has a very centralized institutional structure, where presidential decrees are an important decision-making mechanism, and where federal laws that are developed and adopted in the Duma often are absorbed by President Putin, thus disabling the Duma's veto player role. Subnational government institutions have only limited opportunities to oppose or modify central directives, except in the implementation phase. Although Japan is a democracy, its political institutional structure is also centralized and hierarchical in its design, with a parliamentary system dominated by one political party that is closely aligned with powerful business interests and with weak traditions for veto players at the subnational governance level. India's political system is, on the other hand, characterized by centralized power at the federal level, but has substantial leeway in the implementation phase for state and local governments to oppose, adjust and delay.

Table 9.3 Political parameters

Actors	Political System (Formal Institutions)	Distribution of Political Power (Overall Pattern)
USA	Democratic; federal; pluralist	Multiple veto players enhance status quo stability
EU	Democratic; supranational features but wide member state autonomy	Multiple veto players but consensus-oriented package deal processes enhance mitigation policy
Japan	Democratic; parliamentary; unitary	A moderate number of veto players but sufficient for relatively high status quo stability
Russia	De facto authoritarian; strong presidential power; centralized policy-making	(Very) few but highly effective veto players
Brazil	Democratic; federal	Multiple veto players but policy change can be achieved through coalition building
China	Authoritarian; centralized policy-making	Few veto players, mainly within the central leadership
India	Democratic; parliamentary; moderately centralized	Moderate number of veto players but broad consensus on development imperatives severely constrains mitigation policy

The effects of these political parameters on prospects for policy change seem to depend on other factors, notably demand and supply. In our case studies we find no clear bivariate connection between a high degree of centralization and prospects for change towards more ambitious climate policies. While previous research focusing on Western democracies finds that parliamentary systems tend to enact more climate policies than do federal systems because of fewer veto points (Lachapelle and Paterson,

2013), new research on the Russian policy-making process shows that even if decision-making is highly centralized in Russia, *implementation* of policy is not. It is primarily at the implementation level that veto players in Russia can operate (Korppoo, forthcoming, 2015). This observation is valid also for other countries with a centralized policy-making system, like China and India.

Our case studies of countries with less centralized governance systems – the United States, the European Union and Brazil – clearly emphasize that several potential veto players at different stages of the policy-making process have influenced climate policy trajectories. For instance, in the United States, the executive, legislative and judicial branches of government at the federal and state levels all play different and important roles in policy-making, and have clearly defined possibilities to delay or hinder policy change. This differentiation of powers has allowed opposing interests in the climate and energy policy field to veto policy change. In the European Union, while formal decision-making power is divided between the Commission, Council and Parliament, the process is consensus oriented, often developing package deals that provide joint gains satisfying even potential veto players. Consequently, the European Union has been able to adopt significant changes in the climate policy field by unanimity. Brazil's political system has a multilevel design, with strong presidential powers that depend on coalition building in the Congress. This design allows both the president and coalition partners veto power.

Democracies have institutional structures that create veto players and coalitions that decide the direction and ambition of climate policy change. In non-democracies, we found that key agencies or elites within the governmental structure play important roles in deliberations about policy change. These roles have some of the same effects as do those of veto players in democracies. For instance, both the federal structure of the US political system and the autocratic structure of China's political system produce veto players. In the United States, however, there are more veto players at several governance levels than there are in China. The degree of centralization and of fragmentation of decision-making power is important for understanding the dynamics of the policy-making processes in both countries, but the specific effects also depend on the two countries' respective modes of governance.

In sum, the assumption that more centralization of powers in itself facilitates policy change did not find much support in our case studies. However, as the next section will show, centralization and governmental supply interact and their *combined* effect can be strong.

Driving Forces: Governmental Supply

In Chapter 1, we expected active supply of new governmental initiatives to increase the likelihood of climate policy change. We assumed that the policies pursued by government would depend on the preferences and beliefs of governmental actors, the distribution of authority and power among these actors, and the extent to which government controls the activities to be regulated. In general, we expected that if multiple governmental actors have jurisdiction over a certain policy domain or are otherwise empowered, the process will be more complicated than if only a small number of actors have power to propose new policies. We expected to find clear differences between democracies and non-democracies in this respect, with clearer specification and differentiation of powers in democracies than in autocracies. Moreover, we expected developing countries to see economic growth as a political imperative, leaving the responsibility for climate change mitigation largely with the rich North.

In the case studies we found that horizontal fragmentation and vertical decentralization also help explain why the strength and form of governmental supply of new climate policy initiatives make a difference. In a centralized political system, for instance, strong push from the government towards policy change is essential for success. Consider the difference between Russia and China. In Russia, strong presidential powers have not been applied to supply new climate policy initiatives, and no policy change in the direction of more low-carbon initiatives has occurred. In China, on the other hand, strong presidential powers laid the foundation for the greater prioritization of low-carbon policies since 2007. This increased prioritization may appear surprising given our expectation that developing countries' preferences for strong climate change mitigation policy will have to yield to a much stronger imperative to foster economic growth and to eradicate poverty. China's drive for economic growth does, however, create a rapidly growing need for energy, and concerns with energy security speak in favour of diversifying supply. In democratic developing countries with emerging economies, such as India and Brazil, such governmental priorities are clearly restricting the options for governmental supply. This effect is especially visible regarding fossil energy subsidies. Removal of fossil energy subsidies is almost impossible because of the immediate effect on citizens' daily expenses; their removal would likely cause social unrest and punishment by the electorate at the next election. Moreover, incentive structures like feed-in tariffs are often too expensive for the government to propose because they require upfront funding. Without such incentives – or significantly lower market prices – it is extremely hard to change the subsidy structure in the energy sector.

Our case studies revealed some interesting differences in the role that agenda setting plays in determining the strength and form of governmental supply. In pluralistic political systems like those of the European Union and United States, with multilevel or federal governance structures, there is a crucial difference in terms of who the agenda setter is (Skjærseth et al., 2013). In the European Union, it is the Commission who introduces new law proposals and can control the initiation and timing for steering a policy proposal through the process. With a unified Commission that agrees on a coherent policy proposal, including the positions of both DG Environment and DG Energy, the supply of policy follows a consensus-seeking route. In the United States, agenda-setting powers are shared between the executive and the legislative branches, and in Congress each member has the right to propose new laws. Congress is the chief law-making institution with independent powers. The president has only limited powers to take executive action without congressional involvement, and Congress can stop the president's initiatives, for instance by refusing funding. In reality therefore, the policy supply process is more straightforward and consensus seeking in the European Union than in the United States. Japan, in contrast to the United States, has a strongly supply-driven process. Japan's Cabinet is the chief formal agenda setter, and the strong dominance of one party, the LDP, has clearly shaped the direction and ambition of climate policy. The LDP has strong ties with certain ministries (METI and MOFA) and the business organization Keidanren, and together these actors quite effectively control climate policy supply in Japan.

In Russia, China and Brazil, powerful elites largely control agenda setting and the ability to form majority coalitions within political institutions. Coalition building tends to happen through horse trading between elites on political issues that they each hold dear. For instance, value conservatives in the Brazilian Congress tend to trade their vote on forest and climate policy in exchange for support on religious issues. In Russia, the president maintains tight relations with an inner circle of powerful domestic actors that provide advice on policy direction. Climate policy that leads to significant costs has not yet made it to the governmental agenda in Russia. In China, the top leadership has strong powers to set the policy-making agenda and to steer new policies through law-making institutions. However, the leadership depends on close relations with and accurate information from key agencies such as the NDRC to shape the policy agenda. In India, there is cross-partisan agreement that any climate action that restrains economic development is unacceptable. Government initiatives clearly prioritize infrastructure building to use domestic and imported coal to fuel economic growth at the expense of climate change mitigation goals.

In sum, the case studies support the assumption that supply-side mechanisms often make an important difference. Moreover, we see that where many governmental actors are empowered or have jurisdiction over a policy domain, the policy-making process will be more complicated and less likely to result in policy change.

Driving Forces: Societal Demand

Our final assumption in Chapter 1 concerned societal demand, where we anticipated that the strength and form of societal demand would influence the ambitiousness and likelihood of climate policy change. In democracies we expected societal demand to be more clearly articulated by a wider range of actors than in authoritarian systems. However, we also expected that even autocrats and ruling party elites would be concerned about public dissatisfaction that could threaten their power base. Moreover, we expected policy change to be more likely and significant in countries where advocacy coalitions can outweigh or escape veto players.

For understanding how social demand constrains or facilitates climate policy change, we needed to look into the different ways that public opinion pressure works. This analysis allows a focus on identifying veto players and the composition of winning coalitions, and identifying crucial differences in veto player mechanisms and coalition-building efforts. A general finding is that all governments, in developed as well as developing countries, experienced a strong and consistent demand for low energy prices. This demand constrains energy-related mitigation measures. There are, however, also clear differences between rich and poor countries. In emerging economies like Brazil, China and India, but also in Russia, the most pressing societal demand is for economic growth that can improve living conditions and provide basic welfare services to the people. India especially has huge challenges in terms of eradicating poverty and improving living conditions. Access to energy at a low price is an essential component of the growth strategies in these countries, and people have very strong incentives to maintain energy subsidies that help provide affordable electricity and gasoline. Even in the wealthier countries in our study, we find that public pressure to keep energy prices low is closely linked to the important role that energy plays in economic growth.

Another finding from our case studies is that strong public support for ambitious climate policy is lacking in most of the actors. In the United States, public opinion regarding the seriousness and urgency of the climate change issue is split along geographical as well ideological cleavages, providing reluctant politicians with sufficient support to block ambitious mitigation initiatives. A very stable coalition of fossil fuel industry interests and coal state politicians has dominated policy-making processes in the US Congress, creating strong political resistance to climate policy action. This coalition will also be crucial for whether President Obama's new climate policy – the Climate Action Plan – can be implemented in time before he leaves office. An important characteristic of this coalition is that it represents a clear priority among constituents for low energy prices. In Russia, the coalition that opposes climate action is also very stable, consisting of the elite within the oil and gas industry that wield strong political influence within the de facto autocratic regime of President Putin. Public pressure for mitigation measures is almost non-existent in Russia. Because policies that can deliver emission reductions have remained inefficient and do not cover the major emitters – the energy and industry sectors – the link between fossil fuel industries and domestic action is weak in Russia. Further, the general short-term approach to policy-making in Russia, combined with scepticism that a new and effective international agreement will be established, reduces public worries over the impacts of climate policies on Russia's fossil fuel exports. In Japan, Keidanren, METI and LDP have served as veto players, and in effect have decided the direction of Japan's climate and energy policy. Public engagement in climate change issues has been weak in Japan, and a well-functioning advocacy coalition is lacking. In China, some ministries – especially the NDRC and other key ministries – have served as important veto players but they have also to a certain extent been counteracted by the coalition of scientists and ENGOs that have managed to put pressure on government to enact climate policy. The Chinese government has made climate change and air pollution mitigation national priorities, and the ministries and local governments are administering their respective areas and looking out for their own interests within this context. There is an increasing awareness about and explicit social demand for clean air, factors that both have prompted local governments to implement policies to better the air quality. In India, there is a stronger focus on adaptation than on mitigation policy. The young, educated segment of the population is generally more concerned about the effects of climate change than other segments are, but adaptation is their priority. Mitigation policies have little support as long as economic growth remains an overarching political imperative.

In the European Union and Brazil, public support for climate action has traditionally been quite strong. In the European Union, however, stark differences exist between member states. Poland and other Central and Eastern European states (CEES) acted as veto players that tried to counteract the climate policy proposals originating in Western and Northern Europe. The European Union barely managed to satisfy these countries through a process of re-packing policies. The re-packaging gave CEES significant concessions, and helped persuade them to agree to the new Climate and Energy Policy framework for 2030. In Brazil, sector interests diverge. The rural caucus acted as a veto player in the forest sector for many years, but had to accept new policies to reduce deforestation after great domestic political pressure about and high international attention to Brazil's deforestation rates. In contrast, strong interest groups in the energy sector with significant clout in the policy-making process support further growth and expansion of the domestic petroleum industry, and actively oppose stronger climate policy action.

In all the countries that we studied, local and more immediate pollution issues trump climate change mitigation on the public's priority list. Since air pollution and other local forms of environmental degradation have harmful effects here and now, they have a stronger immediate political saliency than do the more long-term effects of climate change. Politicians must pay attention to problems their citizens care about, and our case studies show that such responsiveness is an important determinant for policy change in both democracies and non-democracies.

For instance, air pollution and local pollution concerns play an increasingly important role for climate policy in China, India and the European Union. Heavy smog in Beijing and New Delhi causes protest movements and public pressure to alleviate the situation, both of which force the government to seek a replacement for coal as the main energy source. In fact, air pollution mitigation is currently the most effective driver for GHG mitigating policies and energy transformation in China. In the European Union, air pollution remains a concern for public health and the environment and the EU is still far from achieving its long-term objective 'to achieve levels of air quality that do not result in unacceptable impacts on, and risks to, human health and the environment' (European Environmental Bureau, 2014). In the United States, Japan and Russia the concerns about local air and water pollution resonate more strongly with a broader part of the public than climate change concerns do. To the extent that such concerns lead to phase-out of fossil fuels, and coal in particular, the effect will be positive also on CO_2 emission trajectories.

A bullet point summary of conclusions regarding supply–demand impact is offered in Table 9.4. Three observations stand out. First, with partial exceptions for the European Union and Brazil, public demand for increasingly ambitious climate change mitigation policies is weak or at best moderate. Overall, public demand for economic growth and low energy prices trumps demand for climate change mitigation, particularly in developing countries. Moreover, local pollution problems tend to be perceived as more urgent than damage caused by climate change. Second, mitigation policy developments seem to be determined more by supply-side mechanisms than by societal demand. This pattern is seen most clearly where societal demand is weak and the political system authoritarian. Third, even in authoritarian systems, governments monitor public demand and try to respond to widespread dissatisfaction that may threaten their power positions or undermine the regime at large. When much is at stake, authoritarian responses may well be repressive but in the climate change domain, some degree of accommodation seems to be the preferred option.

Table 9.4 Societal demand and governmental supply of climate change mitigation efforts

Actor	Societal Demand	Governmental Supply
USA	Low but with some active pushers	Low (Congress) to high (current president)
EU	Varying, higher in Western than in Eastern Europe	High, embedded in comprehensive package deals
Japan	Low–moderate	Moderate–low, weakened by the Fukushima accident
Russia	Very low	Very low
Brazil	Moderate–high	High but declining
China	Low but growing (due mainly to concern with local pollution)	Moderate but with increasing efforts to control local pollution
India	Low	Low

Note: Societal demand and governmental supply are measured in relative terms and on a five-level ordinal scale ranging from very low to very high.

CONCLUSION

We conclude by briefly addressing three questions. First, what have we learned about the usefulness of the conceptual framework that has guided the empirical case studies? Second, what are the main conclusions about likely short- to mid-term policy trajectories that can be drawn from these studies? Third, what are the main implications of these findings for the prospects of significant progress in the United Nations Framework Convention on Climate Change (UNFCCC) negotiations?

Conceptual Framework

What we outlined in Chapter 1 was a conceptual framework, not a mathematical formula. The main difference between the two is that while both can serve important heuristic functions only the formula can provide precise numerical conclusions. This difference has important implications with regard to criteria for evaluating frameworks and formulas, and the ways in which we have used our conceptual framework and the I = PAT formula in this chapter illustrate the point.

Our comparative assessment of the case studies shows that our framework proved useful in identifying and categorizing factors that have in fact influenced actors' climate policy trajectories. For example, the concept of veto players, precisely defined, proved useful in identifying actors capable of blocking a certain policy option in a given context. Similarly, the material parameters concept – more specifically highlighting domestic energy resource endowments and large-scale material infrastructure lock-in – served as a useful tool for identifying important determinants of mitigation costs and coalition building. In identifying these and other factors as important we relied on insights from previous research to specify their causal significance, at least in terms of the *direction* and preferably also the *strength* of impact. Although sometimes building on statistical data, our assessments of the strength of impact on policy trajectories typically have the format of ordinal-level statements with a small number of categories (two to five).

Both of these operations – identifying important factors and specifying their causal significance – are in functional terms equivalent to those performed in developing a formula (such as I = PAT). Where the latter differs the most from the conceptual framework that we have used to guide our case studies is when it comes to *integrating* factors into one single statement that enables the user to calculate aggregate impact in precise numerical terms. As intellectual construct the formula approach is the more elegant. For policy-makers and stakeholders, however, more complex conceptual frameworks will likely come closer to the kind of challenges that they are facing – particularly when their main challenge is to *decouple* emission trajectories from those derived from the formula.

Likely Policy Trajectories

Over the past two decades, GHG emissions for the world at large have stayed close to the I = PAT trajectory. This trajectory fails, by a wide margin, to achieve the UNFCCC objective of 'preventing dangerous anthropocentric interference with the climate system' (Article 2). Given the political imperative to improve living conditions for hundreds of millions of poor people, meeting the UNFCCC objective will require a *decoupling* of population growth and rising income levels from increasing GHG emissions. The bad news is that for the world at large such decoupling is still a remote prospect.

The good news is that among our seven actors four have recently succeeded in decoupling I = PAT scores from GHG emissions: Brazil, the European Union, Japan and the United States. In all four cases, the decoupling seems to have been achieved through some combination of technological development, government policies and voluntary measures taken by non-governmental actors (including civil society as well as industry and business). Government policy seems a more important driver of change for Brazil and the European Union than for the United States and arguably also for Japan.

Looking towards 2020, we note that in the case of Brazil some of the fairly radical measures of forest protection are now being modified, indicating that most of the emission reduction gains from that source may already have been harvested. Oil production is increasing and CO_2 emissions can be expected to increase (from a relatively low level). The European Union has already adopted policy plans of a more ambitious nature and we expect incremental progress in reducing GHG emissions to continue, although at varying speeds in member states. The Fukushima accident led Japan to close all of its nuclear plants, and with imported fossil fuels as the main source of instant compensation, we see a moderate increase in CO_2 emissions. We also see growing investments in domestic renewable energy, and 'safe' nuclear plants will likely increase production, so emissions levels may be expected to stabilize and perhaps even incrementally decline by 2020. In the United States, climate change mitigation remains a strongly contested issue, and the most optimistic prediction for a federal climate policy seems to be that some of President Obama's recent executive initiatives will survive and be implemented. At the subnational level, states, cities and other actors will continue pursuing divergent policies and achieving different results. Overall, the substantial decline in US CO_2 emissions achieved since 2009 due to the switch from coal to natural gas in the power sector is likely to taper off. We expect Russia to continue to be heavily reliant on fossil fuels. But Russia's economy is facing increasingly volatile times, so no clear trend can be predicted over the next five years with regard to GHG emissions. India will continue insisting on its right – and, in fact, duty – to prioritize economic and human development. Only if emissions cuts can be combined with economic growth will India be positive to change its climate policy course. Investments in renewable energy will likely increase but not enough to move India off its current trajectory that closely follows the I = PAT prediction. Finally, as we have seen, China invests massively in renewable as well as fossil sources of energy. The share of renewables in China's energy mix will likely continue growing. Overall, we expect China's GHG emissions to continue increasing but at a progressively lower rate than that predicted by the I = PAT formula.

Prospects for the UNFCCC Negotiations

Finally, what are the main implications of the findings reported in this book for the prospects of significant progress in the UNFCCC negotiations, looking also beyond the 21st Conference of the Parties in December 2015? A good place to start is to recognize one of the basic limitations of global conference diplomacy: the more demanding the project, the more critical becomes broad participation and/or large contributions from participating parties. One implication of this 'law' is that for a truly demanding mitigation agreement to be implemented by 2020 each of our seven actors – possibly with the exception of Japan and Russia – will be *pivotal* parties, directly or indirectly.[5] This is basically bad news since only few of these parties are willing and able to cut their own emissions. The flip side of the coin is that, acting together, the group would constitute a *winning coalition*, meaning that any agreement signed (and, if necessary, ratified) by all seven will almost certainly be accepted by an overwhelming majority of other countries. In fact, since the group controls more than 70 per cent of total GHG emissions it would be in a position to achieve significant cuts without contributions from others. Also a subgroup bridging the North–South divide – notably, one including the EU, the USA, China and India or Brazil – could likely attract sufficient support to reach a new and more effective global climate change agreement.

For either of these coalitions to emerge, leadership by one or more frontrunners will be required. Two recent articles examined the prospects of such leadership to emerge. Underdal et al. (2012) explored whether and to what extent a *conditional* commitment by a major actor (the EU was used as one example) could induce other major actors (such as the USA, China, India or Japan) to enhance their mitigation efforts. The authors found that emission cuts by any of these five actors would reduce mitigation costs for all the others and thereby expand the settlement range for a new climate agreement. However, these mitigation costs reductions were found to be insufficient to induce *significant* policy change. In a forthcoming paper (Hovi et al.), agent-based modelling is used to explore under which condition (if any) two or more frontrunners can succeed in establishing a small 'climate club' and attract new members through club goods (i.e., benefits reserved for club members only) and/or conditional commitments to enhance mitigation efforts. The analysis shows that even a club with less than a handful of major actors as initial members *can* grow and eventually reduce global emissions significantly, but only under a set of 'favourable' conditions. Conditions that can induce club growth include the provision of large incentives from frontrunners to (initially) reluctant countries, and no 'contamination' of negotiations from conflicts over other issues. There are few indications in current international climate negotiations that such conditions will occur in the short to medium term.

Coupling findings from these two papers with our analysis of near-term policy trajectories we conclude that both types of coalitions – the all-seven coalition as well as the smaller subgroup coalition – could make a real difference. However, both would likely fail by fairly wide margins, at least in the short to medium term, to deliver an agreement that is broad and deep enough to accomplish what IPCC sees as prudent to achieve the 'ultimate objective' of the UNFCCC regime, namely to 'prevent dangerous anthropogenic interference with the climate system'.

NOTES

* We would like to thank all contributing authors in this volume for valuable input to the comparative analysis and conclusion.

1. *** indicates that the correlation is significant at the 0.001 level; ** indicates significance at the 0.01 level; * indicates significance at the 0.05 level.
2. Here, 'decoupling' refers to the combination of increasing I = PAT scores with declining GHG emissions.
3. The European Union's I = PAT score increased by about 70 per cent from 1990 to 2011, while total GHG emissions decreased by 15 per cent.
4. During this six-year period, Brazil's I = PAT score increased by 45 per cent.
5. A party is *directly* pivotal if its own contribution is necessary to accomplishing the project. A party is *indirectly* pivotal if it leads or speaks for a larger group whose *aggregate* contribution is necessary for the project.

REFERENCES

Batidzirai, B., E.M.W. Smeets and A.P.C. Faaji (2012), 'Harmonising bioenergy resource potentials – methodological lessons from review of state of the art bioenergy potential assessments', *Renewable and Sustainable Energy Reviews*, **16**(9), 6598–630.

Bo, X. (2014), 'China unveils energy strategy, targets for 2020', *Xinhuanet*, 19 November, accessed 29 March 2015 at http://news.xinhuanet.com/english/china/2014-11/19/c_133801014.htm.

European Environmental Bureau (2014), 'Position paper: revision of the National Emission Ceilings (NEC) Directive', October 2014, accessed 29 March 2015 at http://www.eeb.org/EEB/?LinkServID=9168DFDD-5056-B741-DBA5B64C92B81F9D.

Hovi, J., D.F. Sprinz, H. Sælen and A. Underdal (forthcoming), 'The club approach: a gateway to effective climate cooperation?', paper under review.

Korppoo, A. (forthcoming, 2015), 'Veto players in Russian climate policy: applying and adjusting Tsebelis' theory to a non-democracy', *International Environmental Agreements*.

Lachapelle, E. and M. Paterson (2013), 'Drivers of national climate policy', *Climate Policy*, **13**(5), 547–71.

Lu, X., M.B. McElroy and J. Kiviluoma (2009), 'Global potential for wind-generated electricity', *PNAS*, **106**(27), 10933–8.

Skjærseth, J.B., G. Bang and M. Schreuers (2013), 'Explaining growing climate policy differences in the European Union and the United States', *Global Environmental Politics*, **13**(4), 61–79.

Smeets, E.M.W., A.P.C. Faaji, I.M. Lewandowski and W.C. Turkenburg (2007), 'A bottom-up assessment and review of global bio-energy potentials to 2050', *Progress in Energy and Combustion Science*, **33**, 56–106.

Timilsina, G.R., L. Kurdegelashvili and P.A. Narbel (2012), 'Solar energy: markets, economics and policies', *Renewable and Sustainable Energy Reviews*, **16**(1), 449–65.

Underdal, A., J. Hovi, S. Kallbekken and T. Skodvin (2012), 'Can conditional commitments break the climate change negotiation deadlock?', *International Political Science Review*, **33**(4), 475–93.

US Energy Information Administration (2011), 'International energy statistics – coal', accessed 1 May 2015 at http://www.eia.gov/cfapps/ipdbproject/IEDIndex3.cfm?tid=1&pid=1&aid=2.

US Energy Information Administration (2012), 'International energy statistics – natural gas', accessed 1 May 2015 at http://www.eia.gov/cfapps/ipdbproject/IEDIndex3.cfm?tid=3&pid=26&aid=2.

World Energy Council (2010), Table 7.1 in *2010 Survey of Energy Resources*, accessed 29 March 2015 at http://www.worldenergy.org/publications/2010/survey-of-energy-resources-2010.

Index

Note: Due to space restrictions, many main entries can also be found as sub-entries under the following countries: Brazil, China, European Union, India, Japan, Russia and United States.